D1268267

DRUG–NUTRIENT INTERACTIONS

In clinical medicine, the importance of considering the effect of one drug on the efficiency or metabolism of another drug has been recognised for some time, but relatively little consideration has been given to the effect of drugs upon nutrition or of nutritional alterations upon drug metabolism. In recent years, however, it has become increasingly obvious that the nutritional status is one of the major factors capable of modifying the pharmacological effect of drugs. On the other hand, drug treatment of disease may impair the nutritional status of a patient, with consequent profound effects upon the body's response to drugs. The question of nutrition, drugs, and their interrelations is therefore of great importance. An understanding of these interrelations may lead to a more rational approach to drug dosage and to a reduction in adverse drug reactions and interactions in order to achieve better use of drugs and better public health in general. Furthermore, a number of naturally occurring carcinogens and mutagens have been identified in foodstuffs. Others can be formed during cooking and processing of foods. Considerable experimental and epidemiological evidence has accumulated suggesting that nutrition is a major influence in modifying the effects of these foreign compounds.

This volume has covered these aspects with appropriate examples. The book should serve as a valuable reference source for clinical nutritionists, dietitians, biochemists, and all concerned with the evaluation of foreign compounds, especially physicians, who are often bewildered by the vast array of drugs now on the market.

T.K. Basu is Professor of Nutrition, University of Alberta, Edmonton, Canada.

DRUG-NUTRIENT INTERACTIONS

T.K. BASU, B.V.Sc., M.Sc., Ph.D., FACN
Professor of Nutrition and Honorary Professor of Medicine
The University of Alberta
Edmonton, Alberta, Canada T6G 2M8

CROOM HELM
London • New York • Sydney

RM302.4
B37
1988

© 1988 T.K. Basu
Croom Helm Ltd, Provident House,
Burrell Row, Beckenham, Kent BR3 1AT
Croom Helm Australia, 44–50 Waterloo Road,
North Ryde, 2113, New South Wales

Published in the USA by
Croom Helm
in association with Methuen, Inc.
29 West 35th Street
New York, NY 10001

British Library Cataloguing in Publication Data

Basu, T.K.
    Drug-nutrient interactions.
    1. Drug-nutrient interactions.
    I. Title
    615'.704   RM302.4
    ISBN 0–7099–3203–0

Library of Congress Cataloging-in-Publication Data

Basu, Tapan Kumar.
    Drug-nutrient interactions.

    Bibliography: p.
    Includes index.
    1. Drug-nutrient interactions. I. Title [DNLM:
1. Drug Therapy — adverse effects. 2. Drugs — metabolism.
3. Metabolism — drug effects. 4. Nutrition — drug
effects.   QV 38 B327d]
RM302.4.B37 1988   615'.73   87-30569
ISBN 0-7099-3203-0

Printed and bound in Great Britain
by Billings & Sons Limited, Worcester.

Contents

OCT 1 1 1988

Foreword

In the world at large and in the Western World in particular, the average age of the population is increasing. This is related to an increase in lifespan resulting from remarkable advances in preventive medicine and the clinical sciences. There has also been a concomitant rise of the modern pharmaceutical and chemical industries which support modern treatment methods and influence the pattern of human disease. The science of nutrition has also made major advances in recent years and is poised for even more encouraging contributions as the tools of molecular biology are applied to mechanisms of nutrient effects at the molecular level.

Instruction in nutritional science can no longer be restricted to a description of the chemistry of major dietary constituents, diseases associated with a deficiency, and the amounts of nutrients required to prevent them. Modern nutritionists must now address the pervasive interrelationships of long-term nutritional habits and chronic diseases of the cardiovascular system, of cancer, and of osteoporosis, among others. There is also the role of nutrition as a tool in the treatment of post-operative and other patients in the clinical setting. It is at these interfaces that drugs and nutrients interact in significant ways.

The advances and applications of modern drug therapy and of nutrition are not unrelated. A well-nourished individual is less likely to contract disease, and when specific diseases are encountered, they tend to be less severe and response to drug therapy more favourable. Moreover, response to drug therapy may differ in the same individual at different time points. A significant proportion of these variances appears to be a reflection of the nutritional status of the individual and the food that is being consumed at the time of drug therapy.

The concept of drug–nutrient interactions is not new, but it has only recently gained currency in medicine and in the treatment of diseases in various segments of our populations, particularly the aged. Drug–nutrient interactions can be defined as events that result from chemical, physical, physiological or pathophysiological relationships between nutrients and drugs. These events are important when they diminish the intended response to a therapeutic drug; when the nutritional status of an individual is impaired; or when it causes an acute or chronic drug toxicity.

The term 'drug' in the context of this volume is applied to chemicals of natural origin, of which there are literally thousands to which humans are exposed, as well as those substances produced by chemists for specific diseases, usually referred to as xenobiotics. This volume covers the interactions between these chemicals and nutrients in an admirable fashion. Dr Basu has captured the excitement of this rapidly emerging field of medicine by combining the basics of mechanisms, where known, with ultimate outcomes and has pointed out how these interactions modify metabolism to the benefit or increased risk of an individual. These insights are then linked to the real or potential role of such interactions in long-term health and in the prevention and treatment of disease.

This volume begins with a carefully compiled account of the nature and origin of non-nutrient substances in the diet. Many of these are not only toxic but some are carcinogenic. Their effects on biological systems can be modified by nutrients and, the toxins can in turn influence the availability or utilisation of nutrients. These naturally-occurring agents as well as man-made agents (drugs) are considered in the introductory chapter, along with the clinical implications of modulating the rate of metabolism, a characteristic feature of some drug–nutrient interactions.

There are then three chapters which consider in detail the fate of nutrient substances in the body and how they affect metabolism of natural and foreign compounds; the consequences of drug therapy to nutritional status including effects of drugs on macro- and micronutrients in biological systems; and, specifically, the highly significant area of alcohol and nutrition.

It has become clear in recent years that unbounded enthusiasm for vitamins and consumption of large quantities of them as supplements cannot be practised on a long-term basis with impunity. The unfortunate excessive advertisements in the health foods industry has misled many into believing that if a little is good, a lot is better. There are serious side effects from essential vitamins including Vitamin C, Vitamin A, pyridoxine, and Vitamin D, among others. The author has in Chapter 5 placed these conflicting data and contradictory information in perspective and recommended directions for research to resolve many of the existing questions.

Chapter 6 covers drug–food interactions and discusses data

which clearly show that some foods reduce drug absorption, some increase it. Furthermore, foods affect urinary excretion of drugs and, in other cases, foods can cause adverse reactions.

Chapter 7 covers the subject of nutrition and experimental carcinogenesis. The past decade has witnessed the emergence of a profound interest in diet, nutrition and cancer, an area of considerable controversy and, to some considerable extent, confusion. Dr Basu has treated this in a balanced manner pointing out those cases where the data seem to support a beneficial effect, those where the data are equivocal or contradictory and, instances where the data are unclear or negative. A general set of recommendations, that are provided, regardless of the influence on risk for cancer, cannot help but improve the general health of the public. There are a number of specific recommendations but generic suggestions are to eat a balanced diet, refrain from smoking, and, if alcohol is consumed, do it in moderation. These suggestions may, in the long run, help prevent some forms of human cancer.

Finally, Chapter 8 covers general conclusions which point out that food shortages, as in the developing countries, or the wrong choice of foods in affluent societies contribute to adverse drug–nutrient interactions. Selected examples serve to illustrate the directions nutritionists and physicians should take to prevent or reduce risks for populations as a whole, and particularly, for risk sub-populations such as the elderly who are at higher risk.

For such a wide coverage, the volume is relatively short, easily read, and illustrated extensively with tables and figures. The literature is covered in depth and breadth and the volume will prove to be attractive to those working in nutrition, in toxicology and carcinogenesis, and to the practising physician. It will also prove invaluable to students aspiring to careers in these respective areas, suggesting not only the excitement but the complexity of nutrition, pharmacology and toxicology and how they interact. Dr Basu, with this volume, has made a major contribution to the advancement of knowledge in drug/nutrient interactions and will no doubt provoke the interest and imagination of a broad segment of biologists.

Dr Paul M. Newberne
Mallory Institute of Pathology Foundation
Boston, Massachusetts

Preface

In clinical medicine, the importance of considering the effect of one drug on the efficiency or metabolism of another drug has been recognised for some time, but relatively little consideration has been given to the effect of drugs upon nutrition or of nutritional alterations upon drug metabolism. In recent years, however, it has become increasingly obvious that nutritional status is one of the major factors capable of modifying the pharmacological effect of drugs. On the other hand, drug treatment of disease may impair the nutritional status of a patient, with consequent profound effects upon the body's response to drugs. The question of nutrition, drugs and their interrelations is therefore of great importance. An understanding of these interrelations may lead to a more rational approach to drug dosage and to a reduction in adverse drug reactions and interactions in order to achieve better use of drugs and better public health in general.

Furthermore, a number of naturally occurring carcinogens and mutagens have been identified in foodstuffs. Others can be formed during cooking and processing of foods. Considerable experimental and epidemiological evidence has accumulated suggesting that nutrition is a major influence in modifying the effects of these foreign compounds.

This volume covers the above aspects with appropriate examples. The book should serve as a valuable reference source for clinical nutritionists, dietitians, biochemists, and all concerned with the evaluation of foreign compounds, especially physicians, who are often bewildered by the vast array of drugs now on the market.

T.K. Basu
Edmonton

Dedicated to my supportive family — my wife Sutapa, and my daughters Anita and Malini.

1

Introduction

There are six major components of the diet, namely carbohydrate, protein, fat, vitamins, minerals and water, which are considered to be the essential nutrients. The first three components are necessary to yield energy, maintain growth, and repair tissues subjected to wear and tear. Vitamins, minerals and water, although they do not yield energy, are essential for the utilisation of energy and for the synthesis of various necessary metabolites such as hormones and enzymes. The minerals are also incorporated into the structure of tissues. In solution, they play an important role in acid–base equilibrium.

The environment in which we live contains innumerable non-nutrient chemicals of differing nature and origin. A number of compounds may be introduced into the body either by design or by accident that are not normal constituents of the body. These non-nutrients include food additives, such as preservatives, antioxidants, sweeteners, flavours and colours; therapeutic drugs; cosmetics; detergents; pesticides; and industrial chemicals. These substances are 'foreign' to the major metabolic pathways of the body concerned with the metabolism of nutrients, and have variously been called 'foreign compounds', 'xenobiotics' or 'anutrients'. In addition to these synthetic foreign compounds, a wide variety of plant foodstuffs contain anutrients belonging to various chemical groups, such as terpenoids, alkaloids, ethers, flavonoids and cyanogenic glycosides. The variety of these natural anutrients that enter the body is much greater than the number of drugs, food additives or insecticides. However, both the naturally and artificially occurring compounds in food are potentially toxic if they are consumed or if they accumulate in the body.

1

The majority of chemicals normally regarded as foreign to the body are metabolised and transformed into other substances, irrespective of whether the foreign chemical is toxic or innocuous (see Chapter 2). In the case of toxic compounds, metabolism can play an important role in reducing or increasing the toxic effects, for a compound may produce signs of poisoning either because it is toxic *per se* or because it is converted in the body into a toxic substance. The extent to which a toxic substance can exert its deleterious effects may therefore depend on how effective the body is in metabolising it. If a compound is toxic *per se* and is transformed in the body to a less toxic or non-toxic metabolite, the more rapidly it is metabolised the less will be the toxic effect of a given dose. On the other hand, if it is metabolised to a more toxic agent, the more rapidly it is metabolised the greater will be its toxic effect.

1.1 NATURE AND ORIGIN OF ANUTRIENTS IN THE DIET

1.1.1 Naturally occurring compounds

A variety of terpenoids are common constituents of essential oils and are used in perfumery and as flavourings in foods and drinks. These are citral, linalool, limonene, carvone, nerolidol, squalene and β-ionone. Citral (3,7-dimethyl-2,6-octadienal) is widely distributed and occurs to an extent of 75–80% in oil of lemon grass, the volatile oil of *Cymbopogon citratus* or of *C. flexuses*; it is also present in the oils of verbena, lemon and orange. Borneol (2-hydroxycamphane-2-camphanol) occurs in the essential oil obtained from *Dryobalanops aromatica* of the Dipterocarpaceae, and many other plants. Linalool (3,7-dimethyl-2,6-octadiene) is the chief constituent of linaloe oil; it also occurs in the oils of Ceylon cinnamon, sassafras, orange and the flowers of many other plants. Limonene (*p*-mentha-1,8-diene or 4-isopropenyl-1-methylcyclohexene) occurs in various oils, particularly in those of lemon, orange, caraway, dill and bergamot. Carvone (*p*-mentha-6,8-dien-2-one) is present in caraway-seed and dill-seed oils. Nerolidol (3,7,11-trimethyl-1,6,10-dodecatrien-3-ol) is present in many plants and, in particular, in the essential oils from flowers, such as *Melaleuca*, members of the Myrtaceae and *Myroxylon*. Squalene

2

(2,6,10,15,19,23-hexamethyl -2,6,10,14,18,22-tetracoshexaene) is present in large quantities in shark-liver oil and in smaller amounts in olive oil, wheatgerm oil, rice-bran oil and yeast. β-Ionone (4-{2,6,6-trimethyl-1-cyclohexen-1-yl}-3-buten-2-one) is present in essential oils of various plants including *Boronia*.

In addition to the terpenoids, we are also exposed to a variety of alkaloids. Thus, caffeine (1,3,7-trimethylxanthene) occurs in tea, coffee, mat leaves, guarana paste and cola nuts; gramine (3-{dimethylaminomethyl} indole) is present in mutants of barley and other plants; arecoline (methyl-1,2,5,6-tetrahydro-1-methylnicotinate) is present in leaves of *Berginea*, members of the Saxifragaceae, blueberry, cranberry and pear. Piperine (1-piperonylpiperidine) is present in black pepper; tyramine (4-hydroxyphenethylamine) and tryptamine (3-{2-aminoethyl} indole) are present in ripe cheese, purified animal tissue and in fungal products such as yeast extract ('Marmite') and ergot.

Other naturally occurring substances are ethers and flavonoids. The etherial compounds vanillin and coumarin are both used as flavouring agents. Vanillin (3-methoxy-4-hydroxy-benzaldehyde) occurs in vanilla essence, potato parings and in Siam benzoin; coumarin (1,2-benzopyrine) is present in sweet clover, tonka beans, lavender oil and woodruff (*Asperula* spp.). Flavonoids occur as pigments in various plants and flowers. Plant tissue also contains a wide variety of organic acids, both as salts and as esters. A mixture of flavonoid glycosides, including rutin and hesperidin, is known as citrin, and this has a vasodilatory action on the peripheral circulation.

In addition to the compounds mentioned above, a number of 'cyanogenic' plants are commonly eaten by man or domestic animals. They include cassava (manioc), yam, maize, sugar cane, sorghum, linseed pulses and the seeds of cycads (Montgomery, 1965). Similarly, nitrile compounds (β-aminopropionitrile) with potent neurotoxic properties have been isolated from *Lathyrus* and *Vicia* seeds, and these are consumed very commonly by people in many developing countries under conditions of economic stress. Many sulphur-containing compounds are present in vegetables. These include alkyl sulphide, $(CH_2(=)CHCH_2)S$, in onions and garlic; dithiolisobutyric acid, $(HSCH_2)_2CHCOOH$, in asparagus; and the methyl ester of S-methylthiopropionic acid, $CH_3SCH_2COOCH_3$, in pineapple. These compounds are all potentially toxic if they accumulate in the body.

1.1.2 Artificially occurring compounds

A variety of anutrients (Table 1.1) are added to foods by design, for economic, technological and cosmetic reasons. These substances are introduced to foods as food additives in order to prevent spoilage and to prolong shelf life of food; to facilitate its processing; and to render food attractive to the consumer. Food additives have certainly made a major contribution towards improving nutritional health in both developed and developing countries, as well as controlling food-borne infections. Nitrite is an example, and is unique in meat processing in that it is responsible for enhancing the flavour and colour of cured meats and it protects against *Clostridium botulinum* as well as oxidative changes in lipid-containing foods. The use of this preservative has, indeed, made many potentially hazardous products quite safe. The food-borne infections usually arise from unhygienic practices in the home, in canteens, or in restaurant kitchens, and it is nowadays extremely rare for infections to originate in processed foods.

Table 1.1: Food additives other than nutrients

Groups	Examples
Preservatives	Benzoic acid; sulphur dioxide; sodium nitrite, sorbic acid; propionic acid
Antioxidants	Butylated hydroxyanisole (BHA); butylated hydroxytoluene (BHT); alkyl gallates; ethoxyquin
Emulsifiers	Sodium alginate; carboxymethylcellulose; lecithin
Solvents	Propylene glycol; hexylene glycol
Flavourings	Sodium glutamate; coumarin; vanillin; esters; quinine
Sweeteners	Cyclamates; saccharin; aspartame
Colourants	Azo and fluorescein dyes
Texture improvers	Bicarbonates, phosphates, agar

However, food additives can neither be condemned nor condoned as a class, but each case must be judged upon its merits. It is therefore pertinent to ascertain the benefits as well as the associated risks accruing from the application of food additives. The risk/benefit analysis should be the basis of rational decisions as to which additive should be permitted.

Furthermore, one should bear in mind that food additives may be abused. This applies particularly when faulty processing and handling techniques are disguised, when the consumer is deceived into thinking that food is of better quality than it actually is, and when the result is a substantial reduction in the nutritive value of a food. In this latter context, the use of preservatives, such as sulphur dioxide for example, destroys thiamin in food.

1.1.3 Food contaminants

The widespread use of a vast number and variety of chemicals in industry has led to increasing contamination of the environment with these chemicals. Modern agricultural chemicals, such as pesticides, herbicides and fungicides (Table 1.2), are all complex organic compounds, and are often toxic to animals and human beings. Contamination of the food and environment by these chemicals has certainly provoked much public concern in recent years. One should, however, bear in mind that, without the use of pesticides to protect crops from spoilage by insect pests and fungal diseases, and to protect harvested fruits, cereals and vegetables from the ravages of insects and rodents, the present high agricultural yields would be greatly reduced with a consequent widespread shortage of foods and higher prices. Furthermore, the chemicals are usually applied to the plants before the edible part has appeared or at least sufficiently long before harvesting to permit their removal by rain. There are, however, some compounds which because of their long lives, may remain in the soil or in natural water for a number of years. For example, five years after an application of DDT at a concentration of 0.02 ppm to the waters of Clear Lake,

Table 1.2: Food contaminants

Groups	Examples
Pesticides	DDT; malathion; dieldrin
Herbicides	DNOC (dinitroorthocresol); 2,4-D (2,4-dichlorophenoxyacetic acid)
Fungicides	Biphenyl; dehydroacetic acid
Growth improvers for livestock	Antibiotics; oestrogens; phenylarsonic acids
Packaging contaminants	Organotins, butylphthalate, heavy metals, lubricants

California, fish from the lake showed a concentration of 250–400 ppm, mostly in their lipid tissues. Hence, the prediction of safe concentrations of agricultural chemicals becomes extremely difficult because of the tendency of animals and plants to concentrate these substances, especially the chlorinated hydrocarbons.

1.2 POTENTIAL CARCINOGENS IN FOODS

It has been suggested that 80–90% of all human cancers are causatively related to environmental factors (Wynder and Gori, 1977), and that a variety of foreign compounds entering the body via food, play a dominant role in the pathogenesis of the disease, especially hormone-related cancer and cancer of the digestive tract (Armstrong and Doll, 1975; Wynder and Gori, 1977; Fairweather, 1981). Much of the evidence, however, comes from epidemiology, migrant studies, and animal experimentation. It is therefore pertinent to point out that the associations between dietary foreign chemicals and cancer must be interpreted with caution.

Numerous food-borne chemicals with a wide range of structures are suspected to be carcinogens in humans as well as in experimental animals (Miller and Miller, 1976). These are polycyclic hydrocarbons, polycyclic amines, aminoazobenzene derivatives, aliphatic dialkyl-nitrosamines, and other aliphatic alkylating agents, such as epoxides, lactones, ethyleneimines, and the nitrogen and sulphur mustards. The carcinogenic chemicals also include inorganic compounds such as Ni^{2+}, Be^{2+}, Cd^{2+} and Co^{2+} compounds, chromates and certain silicates.

1.2.1 Naturally occurring chemical carcinogens

Not all potential food-borne carcinogens are contaminants or additives. There are a variety of naturally occurring chemical carcinogens, most of which are metabolites of green plants and fungi. Aflatoxins are mycotoxins, which are the products of the fungus *Aspergillus flavus*. A variety of mycotoxins have been identified, which include aflatoxins B_1, B_2, G_1 and G_2. All these compounds are known to be carcinogenic in a wide range of species; however, the most potent hepatocarcinogenic effect in experimental animals appears to relate to aflatoxin B_1 (Wogan, 1973). As much as 10% of a dose of aflatoxin B_1 can be converted to protein- and nucleic-acid-bound forms in rat liver, where it remains as aflatoxin B_1–2,3-oxide, which may be a possible ultimate carcinogenic form (Swenson *et al.*, 1974). Food contamination with aflatoxin may be hazardous in man, since primates have also been shown to be susceptible to its toxic effect (IARC, 1972). Furthermore, aflatoxin is present in peanuts, which are raised as a staple food in southern India, and the incidence of hepatic cirrhosis in that population is high. The incidence of hepatoma has also been found to be very high in Indonesia where fermented peanut cake is commonly used, which appears to contain aflatoxin (IARC, 1972; Peers and Linsell, 1973). The conditions that favour the formation of aflatoxins also favour the formation of another mycotoxin by *Aspergillus* species. The mycotoxin is called sterigmatocystin, which seems to possess one-tenth the hepatocarcinogenic activity of aflatoxin B_1 (Van der Watt, 1974).

Cycasin (methylazoxymethanol-β-glucoside) is a naturally occurring substance present in the palm-like cycad trees (*Cycas circinalis*), which are used as a source of starch in some Pacific islands. When administered orally, cycasin is highly carcinogenic to the liver and kidney of rats and other species (Laqueur and Spatz, 1968). Intestinal bacteria contain β-glucosidase, which breaks cycasin to methylazoxymethanol. Subsequently, this hydrolysed product is decomposed at neutral pH to an electrophilic intermediate, which tends to methylate nucleic acids and proteins. The pyrrolizidine alkaloids are also of plant origin (from the genera *Senecio, Crotalaria* and *Heliotropium*). They contain an allylic ester structure which is responsible for hepatotoxicity and carcinogenicity. The naturally occurring alkaloids with branched-chain allylic esters are normally more hepatotoxic than those with less complex ester functions (Bull *et al.*, 1968).

Safrole (1-allyl-3,4-methylene dioxybenzene) is found in spices and oils (such as sassafras oil), and is also hepato-carcinogenic following metabolic activation (Miller, 1978), Safrole is hydroxylated to 1-hydroxysafrole, which is a stronger carcinogen than is the parent carcinogen (Borchert *et al.*, 1973). Safrole was used as a flavouring agent in the United States until 1960.

7

1.2.2 Metabolic reactivity of chemical carcinogens

There appears to be substantial evidence showing that only a small number of potential carcinogens initiate their carcinogenic effect by reacting directly with macromolecules such as DNA and protein. The majority, however, require activation to chemically highly reactive metabolites before forming readily covalent bonds with macromolecules (Figure 1.1). Such activations occur through a variety of chemical reactions catalysed by the enzymes localised in the hepatic endoplasmic reticulum (see Chapter 2). Thus, aflatoxin undergoes epoxidation to form a highly reactive molecule (aflatoxin 2,3-epoxide); 2-acetamidefluorene is activated through N-hydroxylation; and dimethylnitrosamine is activated through oxidative demethylation to form monomethylnitrosamine, which is subsequently decomposed non-enzymically to the alkylating carbonium ions or diazomethane. There are also a variety of food-related chemicals which are relatively harmless *per se*, but may be converted to potential carcinogens through metabolism in the presence of suitable endogenous factors. Such chemicals include nitrite and bile acids.

Figure 1.1: General scheme for the possible sequences of the metabolism of precarcinogens

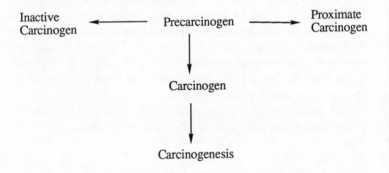

1.2.2.1 Nitrite or nitrate

Nitrates are natural constituents of plants. They are present in vegetables and water supplies. Nitrates may also come from various sources, such as soil, rocks, nitrogen fixation by microorganisms and plants, and nitrate-containing fertilisers, which often favour large accumulation of the compounds in vegetables and water supplies. In addition, nitrates are used as preservatives for fish and meat. Nitrate in our diet could therefore be derived from its deliberate addition to food as an additive, from drinking water, and from vegetables grown in nitrate-rich soils. Nitrate as such is harmless since it is readily excreted in the urine. There is, however, the possibility that nitrate is bacterially reduced to nitrite in food, saliva and achlorhydric stomach. When it happens a hazard arises as nitrite could form carcinogenic nitroso compounds by reacting with secondary amines or amides (Lijinsky and Epstein, 1970). It has been suggested that the intragastric formation of nitroso compounds could be an important factor in the aetiology of gastric cancer (Mirvish, 1971), and that nitrite and a secondary amine or amide must be present at the same time in the stomach to form the compounds (Sander *et al.*, 1968). The necessary amines or amides can be found in any foods, such as fish products, cereals and tea (Lijinsky and Epstein, 1970). They may also be present in drugs (Lijinsky, 1974), which include oxytetracycline, tolazamide, chlorpromazine and quinacrine. For example, methylguanidine, a compound naturally present in fresh beef and fish, has been shown to be converted to a potent carcinogen, methylnitrosourea, after exposure to sodium nitrite in human gastric juice (Endo and Takahashi, 1973).

There is also evidence that nitrosamines can be formed in food products if the concentrations of nitrite and nitrosable amines are high enough and the conditions are appropriate (Sen *et al.*, 1970; Sen, 1972). An epidemiological link between salted fish intake and gastric cancer has been observed (Haenszel *et al.*, 1972). It is possible that the link might be due to the fact that salted fish is a significant source of preformed nitrosamines.

1.2.2.2 Cholesterol and bile acids

Demographic studies (Haenszel and Correa, 1971) have shown that the incidence of colon cancer is related to the *per capita* daily consumption of meat. These observations are in accordance with other studies (Haenszel *et al.*, 1973, 1975), which indicate that high-risk populations consume diets containing a large amount of beef. A substantial socioeconomic gradient of high magnitude in the risk of bowel cancer and socioeconomic differentials in beef consumption have been documented within a single community (Haenszel *et al.*, 1975). In view of the fact that beef contains a relatively large amount of fat which contributes significantly to the overall fat intake in Western communities, it has been suggested that beef fat content rather than beef protein is the principal factor related to large-bowel cancer (Wynder, 1975). Such a view has been substantiated by experimental evidence (Nigro *et al.*, 1975) which has shown that the number of tumours induced by azoxymethane in the small intestine of rats is greater when the animals are given a 35% beef fat diet rather than purina rat chow.

Beef is also rich in cholesterol. It is therefore possible that this steroid is related to cancer of the large bowel. Indeed, inter-country comparisons have shown a high correlation between mortality from large-bowel cancer and that from arteriosclerotic heart disease (Wynder, 1975). There is also evidence (Haenszel *et al.*, 1975) for a substantial social class gradient for myocardial infarction within a community, a finding which is consistent with the differences in the incidence of large-bowel cancer (Table 1.3).

There appears to be substantial evidence indicating that the nature of the diet determines the composition of the intestinal bacterial flora. Thus, examinations of faeces from people of high- (England, Scotland and United States) and low- (Uganda, India and Japan) risk areas have shown that British and American subjects living on refined diets yield significantly more of the gram-negative, non-sporing anaerobic organisms (e.g. *Bacteroides* spp. and *Bifidobacteria* spp.) than do Ugandans, Indians and Japanese living on unrefined diets (Aries *et al.*, 1969; Hill *et al.*, 1971; Finegold *et al.*, 1974). Conversely, the latter group yield more aerobic organisms (e.g. streptococci and enterobacteria), so that the ratio of anaerobic to aerobic is markedly higher in people living on a Western diet than in those on largely vegetarian diets. It is of interest that faecal samples from people in Western countries have also been found to contain higher concentrations of neutral (cholesterol + coprostanol + coprostanone) and acid (tri-, di-, mono- and unsubstituted cholanic acid) steroids than those from people from African and Eastern countries (Hill *et al.*, 1971).

Reddy and Wynder (1973) have investigated the amount and composition of the faecal steroids and bile acids in American

Table 1.3: Social class differences among patients with large-bowel cancer, polyps and myocardial infarction and in *per capita* consumption of beef, pork, eggs and milk in Cali, Colombia

Socioeconomic class[a]	Large-bowel cancer (SIR: 55 years of age and over)[b]	Adenomatous polyps (prevalence/ 100 autopsy specimens; of age and over)	Myocardial infarction (mortality rate/100 000)	*Per capita* consumption of food of animal origin (Ratio: class1/classes III, IV)			
				Beef (1 lb/wk)	Pork (1 lb/wk)	Eggs (No./wk)	Milk (bottles-wk)
	1962–71	1967–69	65 years and over	1964			
I and II	152	33.3	46.1				
III	92	19.8	9.5				
IV	25	3.0	0.7	30	9.3	6.1	5.2

[a] Class 1 (upper; 4.5%) = land owners, managers and some high-income professionals; II (middle; 19.8%) = professionals, businessmen and highly qualified technicians and clerks; III (low; 59.1%) = blue-collar workers and others in low income bracket; IV (very low; 16.6%) = unskilled workers. (Total population of Cali was estimated to be 736 000.)
[b] SIR (standardised incidence ratio) was obtained by application of age-specific incidence rates to the corresponding populations for each group of census tracts to compute expected number of cases (E). The numbers of cases actually observed (O) in each group of census tracts were divided by the expected numbers to calculate the SIR (SIR = O/E × 100).

Source: Data modified from Haenszel *et al.* (1975).

populations with different dietary habits. These include Americans consuming a typical Western diet containing high fat and protein of animal origin, American vegetarians, American Seventh-Day Adventists consuming a largely vegetarian diet, recent Japanese migrants still consuming a more or less typical Japanese diet containing fish and small amounts of meat, milk and eggs, and recent Chinese migrants also consuming a diet low in foods of animal origin. This study has revealed that the faecal excretion of the total neutral steroids is markedly increased in Americans on a mixed Western type of diet compared with that of the other groups consuming a largely vegetarian diet; in contrast the total cholesterol in the faeces of typical Americans is the least. Coprostanol and coprostanone are the breakdown products of cholesterol; the ratio of the sum of the two products to total neutral steroids gives a crude measure of the percentage of cholesterol degraded. This ratio is 0.96 (96%) for Americans on a mixed Western diet, compared with only 0.46 (46%) for Japanese migrants. Thus, although the total neutral steroids in the faeces of typical Americans is much higher than in the other groups, most of it has been converted to coprostanol and coprostanone. A significant increase in the excretion of total bile acids, deoxycholic acid and lithocholic acid has also been found in Americans eating a mixed Western diet.

In addition to the studies of population groups, there have been a number of controlled feeding experiments indicating that it is the fat content of a diet which is primarily responsible for alterations of the gut flora and the faecal steroid composition. In one controlled study (Reddy et al., 1975), eight healthy volunteers living on a high-fat, high-meat Western diet (23% protein, 45% fat, and 32% carbohydrate) were transferred to a non-meat diet, and faecal samples were collected 4 weeks later. The counts of anaerobic microflora together with the faecal excretion of deoxycholic acid and cholesterol metabolites (coprostanol and coprostanone) were higher during the period of consumption of the high-fat, high-meat mixed Western diet compared with the no-meat diet consumption period. In a similar study (Maier et al., 1974), volunteers were fed a high-beef diet for 4 weeks followed by a no-meat diet for another 4 weeks. There was little effect of these diets on the faecal bacterial flora, but the faecal acid and neutral steroid levels fell during the period when no meat was consumed. Other controlled studies (Hill, 1971) in healthy volunteers have revealed that reduction in the amount of dietary fat intake from 100 g to less than 30 g/day results in a rapid reduction in faecal acid steroid concentrations from 6.5 mg to less than 2 mg/g dry weight of faeces by the end of 7 days.

Comparisons between control subjects and patients with colon cancer have shown that the patients tend to have more faecal neutral and acid steroids (Hill *et al.*, 1975; Reddy *et al.*, 1975). Thus the association has been established linking colon cancer to dietary fat and faecal acid and neutral steroids. No dietary carcinogen has so far been isolated that can explain this association. It has been postulated that the interaction between diet and bacteria in the colon of people on a typical Western-type diet produces from certain bile acids a carcinogen or co-carcinogen that acts on colon mucosa.

The faecal excretion of deoxycholic acid and the activity of faecal 7α-dehydroxylase, which converts cholic acid to deoxycholic acid, have been found to be higher in patients with colon cancer than in controls (Table 1.4). Aries and her co-workers

Table 1.4: Faecal neutral steroids and acid steroids in patients with colon cancer

	Total neutral steroids (mg/g dry faeces)	Total bile acids (mg/g dry faeces)
Control (15 patients)	17.8	13.2
Colon cancer (12 patients)	38.4	22.0

Modified data from Reddy *et al.* (1975).

(1969) isolated pure strains of the principal groups from faeces of people from different communities. The ability of the strains to degrade bile salts was tested by incubating washed cell suspensions with their substrate. The bile-degradative enzymes studied were hydrolase, which deconjugates bile salts releasing a free bile acid (cholic acid); 7α-dehydrogenase, which converts cholic acid to a keto derivative; and 7α-dehydroxylase, which, by removing the 7α-hydroxy group, converts cholic acid to a secondary bile acid, deoxycholic acid (Figure 1.2). *In vitro* studies have shown that certain anaerobic bacteria that present in large numbers in the faeces of people living on typical Western diets possess a very high 7α-dehydroxylase activity.

This observation suggests that faecal steroids are much more

Figure 1.2: Bacterial action of bile degradation

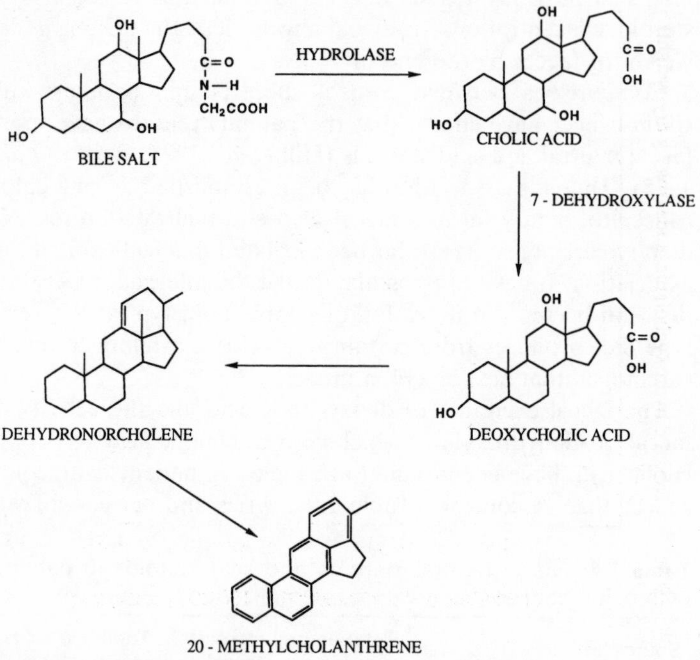

20 - METHYLCHOLANTHRENE

extensively dehydroxylated in people from high-risk areas than in those from low-risk areas. Indeed, there exists a good correlation between the faecal concentration of deoxycholate and the incidence of cancer of the large bowel in six different populations (Figure 1.3).

The secondary bile acid, deoxycholate, has, in fact, been shown to be carcinogenic in experimental animals (Cook *et al.*, 1940). Further, the bile acid is structurally related to a more potent carcinogen, 20-methylcholanthrene (Figure 1.2). There is even some evidence that deoxycholate could be chemically converted into the polycyclic aromatic carcinogen via dehydronorcholene (Haddow, 1958). *In vitro* studies (Hill, 1974) have shown that bacteria can introduce double bonds into the steroid nucleus by four separate nuclear steroid hydrogenase reactions. However, the aromatisation of deoxycholate to 20-methylcholanthrene by intestinal bacteria has yet to be demonstrated.

Experimental evidence (Goldin and Gorbach, 1976) has

Figure 1.3: The relationship between faecal dihydroxycholanic acid concentration and the incidence of cancer of the large bowel

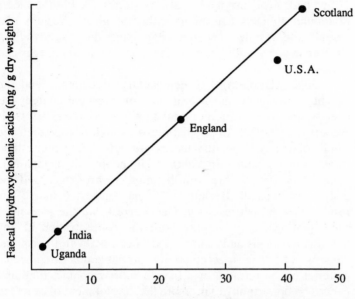

Source: Hill (1971).

revealed that, in addition to 7α-hydroxylase, the activities of other faecal bacterial enzymes such as nitro-reductase, azo-reductase and β-glucuronidase are markedly increased when rats initially fed a diet with a high grain content are transferred to a diet containing mainly beef. These enzymes have been implicated in the conversion of precarcinogens into carcinogens. Thus, nitro- and azo-reductases can convert nitro and azo compounds to aromatic amines which may be carcinogens (Mason and Holtzman, 1975; McCalla *et al.*, 1970). The hydrolytic enzyme β-glucuronidase is believed to be responsible for the regeneration of toxic aglycones through hydrolysis of glucuronides in the lumen of the gut (Reddy *et al.*, 1974). For example, diethylstilboestrol (Fischer *et al.*, 1966) and N-hydroxyfluorenylacetamide (Grantham *et al.*, 1970) are detoxified through glucuronide conjugation in the liver and subsequently excreted via the bile as diethylstilboestrol and N-hydroxyfluorenylacetamide β-glucuronides. The original

15

toxic compounds can be regenerated *in situ* in the bowel under the influence of bacterial β-glucuronidase. An increased activity of this hydrolytic enzyme has also been observed in the faeces of human subjects consuming either a mixed Western diet (Reddy and Wynder, 1973) or a high-meat diet (Reddy *et al.*, 1975), compared with a non-meat regimen containing eggs and milk.

A close relationship between dietary fibre intake and the weight, consistency and transit time of intestinal content has been reported in various population groups (Burkitt, 1971; Burkitt *et al.*, 1972). In populations on a high-residue diet, the stool volume is large, the faeces are soft and the intestinal transit time is short. In contrast, in people living on diets deficient in dietary fibre, transit times are prolonged and the daily stool is small. Burkitt (1975) has suggested that a low residue diet, which results in faecal arrest and the passing of small motions, may increase both the concentration of any faecal carcinogens and their period of contact with the colonic mucosa. If this view is accepted, it is probable that carcinogenic material in the intestine will have a shorter contact time and hence less opportunity to promote the development of cancer in people living on a high-fibre diet.

The addition of dietary fibre in the form of bran has been shown to reduce the degradation of bile salts (Pomare and Heaton, 1973) possibly by inhibiting the 7α-dehydroxylation process. These findings may constitute another possible mechanism whereby dietary fibre could exert a protective effect against the development of cancer of the large bowel.

Experimental evidence (Wattenberg *et al.*, 1962; Wattenberg, 1971), suggests that dietary fibre could also be of importance in determining the response to exposures to polycyclic hydrocarbon carcinogens. Thus, rats fed a purified diet or starved for one day or more have shown almost total loss of benzpyrene hydroxylase activity of the small intestine. However, the enzyme activity is markedly induced when vegetables, particularly those belonging to the Brassicaceae family (including Brussels sprouts, cabbage, turnips, broccoli and cauliflower), are added to the purified diet (purified diet, 75%; dry vegetable powder, 25%).

The production of carcinogens or co-carcinogens by bacterial degradation of bile acids in the large intestine may be important in the aetiology of colo-rectal cancer. It appears from the evidence provided above that the dietary fibre's protective effect against the development of cancer of the bowel may be due to dilution of bile-salt substrate or potentially carcinogenic degradation products. Furthermore, the increased rate of large intestinal transit would allow less time for degradation to occur and for carcinogens to act on the mucosa. The presence of dietary fibre in the large intestine may also alter the physico-chemical environment, which may have a more direct effect on the bacterial flora and its metabolic activity.

1.3 PHARMACOLOGICAL AGENTS (DRUGS)

In addition to the food-borne chemicals mentioned above, drugs are another class (the most important but not the largest) of a group of substances which are chemically foreign to the body and are generally of no nutritive value (Parke, 1968). During the last 50 years, medicine has experienced a profound transformation, with the ability to cure most infectious diseases as a consequence of the discovery of the sulphonamides and antibiotics, the amelioration of mental disease by the use of phenothiazines, tricyclic drugs, barbiturates and diazepines, the successful treatment of cardiovascular disease with diuretics and β-blockers and of gastrointestinal disease with H_2-receptor antagonists and carbenoxolone, and the treatment of many more disease states as a result of the miraculous new discoveries made by medicinal chemists. Unfortunately, however, the interactions between what man eats and the chemicals with which he comes into contact are ever increasing, and yet are so poorly understood. It has become increasingly obvious that nutritional status is one of many factors capable of modifying the pharmacological effect of drugs (see Chapter 3). On the other hand, drug treatment may impair the nutritional status of a patient, with consequent profound effects upon the body's response to drugs. The question of nutrition, drugs and their interrelationships therefore becomes pertinent. Such interactions occur in relation not only to prescribed drugs but also to over-the-counter drugs such as aspirin and antacids. Hence the problem lies with the community at large and involves both healthy individuals and those suffering from disease.

Care must be taken to distinguish between the potency,

toxicity, hazards and pharmacological effects of ingested chemicals, including nutrients. Optimum nutrition requires that intakes of all essential nutrients are in accordance with the recommended dietary allowance (RDA) and that no substances be ingested in quantities large enough to be detrimental to health. However, in recent years, the widespread use of large doses of vitamins and inorganic elements, often by healthy individuals (especially in North America), has become a matter of some concern because of the potential health hazards of this practice. The toxicological and pharmacological aspects of meganutrients are areas that certainly warrant exploration (see Chapter 6).

1.4 CLINICAL IMPLICATIONS OF INDUCED RATE OF METABOLISM OF FOREIGN COMPOUNDS

The ability of a foreign compound to stimulate its own rate of metabolism is of importance in toxicity studies. If the compound produces a metabolite more toxic than the parent drug (see section 1.2.2), then long-term administration may enhance its toxicity. In contrast, if the substance is more toxic than its metabolites, the reverse will be true. The increased activity of the hepatic foreign-compound-metabolising enzymes (see Chapter 2) following treatment with a wide variety of drugs, pesticides, food additives, polycyclic hydrocarbons and other chemicals has been extensively documented (Conney, 1967). The drugs and foreign compounds that induce the hepatic enzymes appear to have widely differing functional properties. For example, phenobarbital is a hypnotic; butylated hydroxytoluene and butylated hydroxyanisole are antioxidants; DDT is a pesticide, and 3-methylcholanthrene is a carcinogen. Furthermore, a variety of indole derivatives (e.g. indole-3-acetonitrile, indole-3-carbinol, and 3,3-diindolylmethane), the constituents of cruciferous vegetables, are known to induce the metabolism of foreign compounds.

There appear to be a considerable number of experimental reports indicating that the induction of the rates of metabolism may result in inhibition of toxicity of a variety of chemicals, especially carcinogens (Wattenberg et al., 1976). The most extensive study of this type has been done with phenolic antioxidants (e.g. BHA and BHT). Thus, BHT, an inducer of hepatic microsomal enzymes, appears to reduce the carcino-genicity of 2-acetamidofluorene and of its more toxic metabolite N-hydroxy-2-acetamidofluorene (Ulland *et al.*, 1973). This protective action of BHT has been shown to result from an increased rate of detoxification of the carcinogens and, in particular, from an increased excretion of glucuronide conju-gates which may result from the induction of the glucuronyl transferases (see Chapter 2). The rapid metabolism may also result in loss of activated carcinogen species due to unavail-ability of critical binding sites to macromolecules such as DNA. Induction of the enzymes deactivating the carcinogens has also been found to protect mice from carcinogens such as benz-pyrene, 7,12-dimethylbenzanthracene and aflatoxin (Watten-berg *et al.*, 1976).

The inhibiting effect of enzyme induction on chemical carcinogenesis depends not only upon the nature of the carcinogen but also upon the relative times of administration of the enzyme inducer and the carcinogen. Thus, simultaneous feeding of rats with phenobarbital (enzyme inducer) and 2-acetamidofluorene (carcinogen) reduces the hepatocarcino-genicity of the latter (Peraino *et al.*, 1971), probably as a result of stimulation of the enzymes detoxifying the carcinogen. In contrast, however, when phenobarbital is administered subse-quent to the feeding of 2-acetamidofluorene, the incidence of hepatomas is significantly increased. Phenobarbital is known to increase DNA synthesis and the proliferation of hepatocytes but the effect is short-lived. It is therefore possible that the enzyme inducer potentiates the carcinogenic activity of the previously administered acetamidofluorene by facilitating the development of cells, malignantly transformed by the meta-bolites of the carcinogen, into established clones.

2

The Fate of Anutrients in the Body

Mammals have developed defence mechanisms that render them capable of metabolising and subsequently excreting a wide range of anutrients. Most drugs in use today for the treatment of human disease owe their activity to a selective toxic effect on an infective or invasive agent, or to the selective inhibition of an enzyme system. The pharmacological action of a drug depends on its effective concentration at its intended target tissue, but the optimum concentration of the active form of the agent will depend on the amount of drug bound by plasma proteins or on the extent to which it is metabolised during its passage through the body (Figure 2.1).

Figure 2.1: General scheme for pharmacological action of drugs

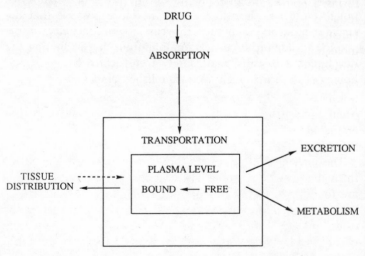

2.1 METABOLISM OF FOREIGN COMPOUNDS

The enzymic breakdown of drugs and other environmental chemicals takes place in two distinct phases. The first phase, biotransformation, is asynthetic and results in the formation of a more polar compound by oxidation, reduction, hydroxylation, deamination, or N-alkylation of a relatively lipophilic anutrient. The second stage is of a synthetic nature, involving conjugation of the polar compound with endogenous molecules such as glucuronic acid, sulphate, glutathione, glycine and acetate, giving the compound more hydrophilic character and thus making it more suitable for excretion in the bile and urine.

The enzyme system responsible for the asynthetic metabolism of anutrients, the mixed function oxidases (MFOs), is principally located in the endoplasmic reticulum of the hepatocytes and is somewhat unique in its ability to utilise a wide range of substrates. The MFO system catalyses the addition of an atom of oxygen into the substrate molecule, converting alkane moieties into phenols and acids, and aromatic hydrocarbons into phenols and epoxides. The membrane-bound enzyme system is complex, and is dependent upon reduced nicotinamide adenine dinucleotide phosphate (NADPH), molecular oxygen, and lipid, and on cytochrome P-450, which acts as the terminal oxygenase (Basu, 1980a). The oxidised cytochrome P-450 present in the endoplasmic reticulum of hepatic cells forms a complex with the drug or anutrient, which is then reduced by NADPH-dependent cytochrome P-450 reductase (Figure 2.2). The reduced cytochrome then combines with molecular oxygen to give an active oxygen complex, and a subsequent breakdown of the oxygen compound produces oxidised cytochrome P-450 and oxidised anutrient. The metabolism of anutrients through the asynthetic pathway may also occur in many tissues other than the liver, especially in the epithelial tissues of the lungs, kidneys, gastrointestinal tract and the skin.

The enzymes responsible for the metabolism of anutrients through synthetic reactions are known as the transferases, and are found in both the endoplasmic reticulum and the cytosol. The most common type of conjugation is the formation of an ester linkage with glucuronic acid. Glucuronide formation occurs *in vivo* with compounds possessing functional groups, such as OH, COOH, NH_2 and SH. Glycine is perhaps the next

Figure 2.2: The microsomal mixed function oxidase system

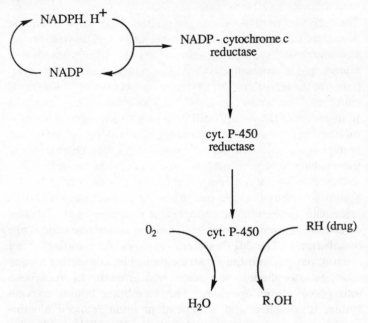

most extensively used conjugating substance. A large variety of aromatic acids and some aliphatic acids are known to be conjugated with the amino acid. The best known glycine conjugation product is hippuric acid (benzoylglycine), which is derived from benzoic acid. Glutamine is another amino acid which can be used as a conjugating agent. One of the known glutamine conjugation products is phenylacetylglutamine, derived from phenylacetic acid. Among other substances which could undergo metabolism through the glutamine conjugation pathway in man are p-aminosalicylic acid and indole-3-acetic acid. A variety of compounds are known to be metabolised through sulphate conjugation. These are primarily aliphatic alcohols (ethylsulphate), secondary cyclic alcohols (androsterone sulphate), amines containing the sulphamyl group (phenyl sulphamate) and phenolic steroids (oestrone sulphate). There are at least five types of amino group which can be metabolised through conjugation with acetic acid, and these are aromatic and aliphatic amino groups, the amino acid group, the hydrazino group, and the sulphonamide group. Methylation is another widespread reaction which involves the transfer of a methyl group from the methionine derivative, S-adenosyl-methionine, to various compounds forming methyl conjugates. Thus, nicotinamide is converted into N-methylnicotinamide, noradrenaline to adrenaline, serotonin to N-methylserotonin, normeperidine to meperidine, histamine to 1-N-methylhist-amine, and pyridine to N-methylpyridine.

By far the most common outcome of metabolism of drugs and other environmental chemicals is deactivation of the com-pounds, thus terminating their pharmacological or deleterious effect. For example, a toxic chemical such as benzene is converted to phenol and subsequently conjugated to phenyl-glucuronide; and benzoic acid is conjugated with glycine and excreted as hippuric acid. The anutrient metabolising enzyme system including both asynthetic and synthetic reactions there-fore acts as a defensive mechanism, protecting the living organism from the hazardous effects of xenobiotic chemicals in the food and the environment.

However, the same enzyme system whose primary function is detoxification can also bring about the converse effect, namely activation of a biologically inert compound to a potentially toxic entity. Thus, bromobenzene is activated through its metabolism into an epoxide intermediate, which can bind covalently with tissue proteins, resulting in hepatic necrosis, and nitrobenzene, which oxidises haemoglobin, producing methaemoglobinaemia. The carcinogens benzopyrine and 2-acetamidofluorene are C- and N-oxgyenated, respectively, to yield active metabolites, which bind covalently to DNA causing mutation and carcino-genesis.

It therefore appears that, in the case of toxic compounds, metabolism can play an important role in reducing or increasing the toxic effects, for a compound may produce signs of poisoning either because it is toxic *per se*, or because it is converted into a toxic substance (Figure 2.3). The extent to which a substance can exert its pharmacological or deleterious effects may therefore depend on how effectively the body metabolises it.

2.2 NUTRITIONAL FACTORS AFFECTING THE METABOLISM OF ANUTRIENTS

The rate at which the metabolic reactions proceed and their

Figure 2.3: Metabolic fates of drugs and other toxic chemicals

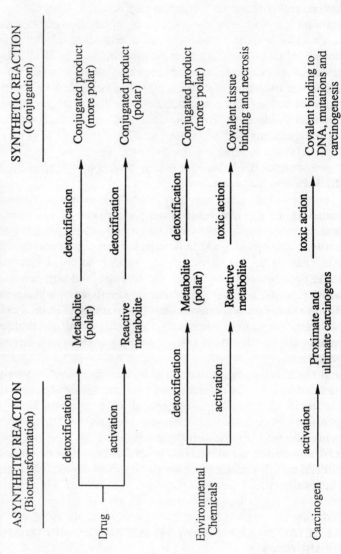

relative importance may be affected by a number of factors, resulting in changes in the pattern of metabolism of a foreign compound and differences in toxicity. These factors are mainly: genetics, age, hormones, disease, exposure to heavy metals, and pre-administration of foreign compounds (Basu, 1980a), but probably the most important factor of all is nutritional status.

The ability of the body to metabolise foreign compounds during concurrent nutrient deficiency has been extensively reviewed (Basu and Dickerson, 1974; Campbell and Hayes, 1976; Ioannides and Parke, 1979; Basu, 1983a). It is evident from these reports that a deficiency in a great variety of macro- and micronutrients affects the metabolism and body disposition of foreign compounds.

2.2.1 Protein malnutrition

Protein energy malnutrition (PEM) due to restricted feeding and starvation is prevalent in much of the world's population. It may occur either as a result of a shortage of food in developing countries or for iatrogenic reasons in affluent communities. PEM has profound effects on antibody production, phagocytic activity, and cell-mediated immunity defence mechanisms, and hence on infectious disease resistance (Nalder *et al.*, 1972; Sellmeyer *et al.*, 1972; Selvaraij and Bhat, 1972). As a result, malnourished subjects become prone to a variety of diseases which are increasingly being treated with drugs; both the drugs and the disease may affect drug metabolism either independently or jointly. Apart from using drugs for treating diseases, the use of pesticides for food production is increasing throughout the world, and perhaps their usefulness is greatest for producing more food in developing countries (Stoewsand *et al.*, 1970) where populations exist on nutritionally inadequate diets.

PEM can affect the metabolism of drugs and other xenobiotic chemicals in at least two ways (Basu, 1977a). First, tissue protein is catabolised and used as a source of energy, thus reducing the availability of amino acids for protein synthesis which, in turn, reduces the amounts of various enzymes in the tissues including those involved in the metabolism of foreign compounds. Secondly, endogenous substrates such as glucuronide, sulphate and glycine are derived from carbohydrate and protein, and there could be competition between the metabolism of anutrients and the needs of the tissues for these nutrients.

Dietary protein alterations are, indeed, known to modify the pharmacological or toxicological activity of a variety of drugs, hepatotoxins and carcinogens through an effect on their metabolism. Protein deficiency has been shown to depress the metabolism of a variety of drugs (Kato et al., 1968). Thus, the activities of hepatic microsomal enzymes that oxidise pentobarbital and strychine, hydroxylate aminopyrine and zoxazolamine, and demethylate aminopyrine are decreased in weanling rats fed either a protein-deficient (5 or 10%) diet for 2 weeks, or a protein-free diet for 4 days, and increased while the animals are fed a high-protein diet (50%). In addition to this, the activities of the NADPH-dependent electron transport systems such as cytochrome P-450 reductase, cytochrome P-450 and cytochrome b_5 have all been shown to be closely related to dietary protein intake (Kato et al., 1968; Marshall and McLean, 1969; Dickerson et al., 1976).

It appears that a protein-deficient diet results in marked depressions of the hepatic microsomal MFO systems. One-quarter of this decrease is due to a depression of hepatocyte proliferation and thus microsomal paucity, and the remaining three-quarters is due to specific enzyme reductions (Campbell and Hayes, 1976). However, in contrast to the effect on MFO systems, feeding animals with a protein-free diet does not appear to impair the synthetic reactions and even results in an increase in the activity of live microsomal glucuronyl transferase (Woodcock and Wood, 1971).

Table 2.1: Effect of high (26%) and low (3.5%) protein diets on the LD_{50} for various pesticides in rats

Pesticide	LD_{50} concentrations (high/low protein ratio, mg/kg body wt)
Captan	26.0
Parathion	7.6
Carbaryl	6.5
Endosulfan	4.2
DDT	2.9
Malathion	2.3
Lindane	2.0
Chlordane	2.0

Source: Adapted from Boyd and Taylor (1969).

In addition to a variety of drugs including pentobarbital, strychnine, aminopyrine and zoxazolamine (Kato *et al.*, 1968), the toxicity of other foreign compounds such as pesticides has been shown to be increased by a protein-deficient diet (Table 2.1). However, protein deficiency does not increase the toxic effects of all foreign compounds, especially those that are activated by metabolism where actually the converse is observed.

McLean and McLean (1966) have shown that in experimental animals a low-protein diet decreases CCl_4 toxicity, suggesting its role of metabolic activation in producing toxic effects. Indeed, it has been implied that CCl_4 may be metabolised, probably to a free radical which may then catalyse chains of auto-oxidation in the structural lipids of microsomes and mitochondria, resulting in fatty accumulation and cellular necrosis (Recknagel and Ghoshal, 1966; Slater, 1966). In young male rats fed a protein-free diet for 7 days, the resistance to the hepatotoxic effects of dimethylnitrosamine seems to be increased (McLean and Verschuuren, 1969). This probably results from the decreased rate of production of methylating metabolites from dimethylnitrosamine by MFO systems (Magee and Lee, 1964).

Heptachlor, a chlorinated hydrocarbon insecticide, is another example of this kind. Feeding a 5% casein diet to weanling male rats for 10 days has been shown to bring about a three-fold increase in tolerance to the insecticide compared with pair-fed mates fed either a 20 or a 40% casein diet (Weatherholtz *et al.*, 1969). The low-protein diet-mediated protective effect against heptachlor toxicity appears to be due to a low production rate of toxic heptachlor epoxide (Weatherholtz and Webb, 1971). In rats fed a low-protein (5%) diet the binding of aflatoxin B_1 metabolites to chromatin, DNA and chromatin protein has been reported to be 70% lower than in rats fed a 20% protein diet (Natarajan *et al.*, 1976). Furthermore, Preston *et al.* (1976) have shown that in paired rats fed a 20% protein diet the binding is slightly less than in those fed a 20% protein diet *ad libitum*.

Although most of the information on the metabolism of foreign compounds in relation to protein nutrition has been derived from animal studies, there is a considerable body of information to suggest that the data derived from animal studies may be applicable to man. Thus the impaired activity of hepatic

27

microsomal MFO systems in children with protein deficiency may be inferred from the findings that they excrete higher amounts of unmetabolised chloroquine (an antimalarial drug) while ill than they do after recovery (Wharton and McChesney, 1970). Furthermore, Mehta and his associates (1975) have shown that malnourishment in children is associated with an increase in plasma half-life and a decrease in the urinary conjugated fraction of chloramphenicol. These results may be attributed to a slower glucuronide conjugation of the drug.

More recently, it has been demonstrated that changes in protein and carbohydrate composition of the diet markedly influence the rates of biotransformation of drugs in man (Alvares *et al.*, 1976a, b; Kappas *et al.*, 1976; Anderson *et al.*, 1982). In these studies, healthy young male volunteers fed a low-carbohydrate high-protein diet for 2 weeks showed a 35–40% decrease in the plasma half-lives of antipyrine and theophylline as compared with the half-lives of these substances in plasma from subjects fed their usual home diets. On the other hand, a change from a low-carbohydrate high-protein diet to a high-carbohydrate low-protein diet for 2 weeks resulted in a 50–60% increase in the half-lives of the two drugs. These changes in half-lives have also been shown to be accompanied by changes in metabolic clearance rates but not in the apparent volumes of distribution.

The relevance of the experimental data to humans is further evidenced by the fact that children with kwashiorkor (protein deficiency) appear to be protected from the hepatotoxic effect of tetrachloroethylene, which is used to treat hookworm infestation (Balmer *et al.*, 1970). It is possible that depression of the hepatic microsomal MFO systems in children with protein deficiency slows the rate of biotransformation of tetrachloroethylene into its toxic metabolite, trichloroacetic acid (Basu, 1980c).

The anti-parkinsonism drug, levodopa, is another agent the pharmacological action of which can also be modified in patients by dietary protein intake. In this case, however, a high-protein diet diminishes and a low-protein diet potentiates the therapeutic effect of the drug (Gillespie *et al.*, 1973; Mena and Cotzias, 1975). Since levodopa is absorbed and transported by the same mechanisms which transport neutral and basic amino acids (Granenus *et al.*, 1971; Richter and Weiner, 1971), it is possible that the high-protein diet-mediated impairment of the pharmacological action of the drug is due to competition between levodopa and the amino acids for their absorption and entrance into the brain.

There appear to be significant fragmentary human data to show that the effects of protein nutrition on the metabolism of drugs and toxic chemicals, which have been derived from studies on experimental animals, may also occur in man. These effects have considerable importance in relation to the biological effects of foreign compounds in individuals suffering from protein malnutrition, in chronically ill patients, in post-operative patients receiving glucose, and in a large segment of the population who manipulate their diets by weight-reducing regimens.

2.2.2 Dietary lipid

Lipids are an integral part of the composition and structure of subcellular membranes. Moreover, in a solubilised and reconstituted system consisting of cytochrome P-450 and cytochrome P-450 reductase, the phospholipid phosphatidylcholine has been shown to be essential for electron transfer and oxygenation of a variety of foreign compounds (Strobel *et al.*, 1970). It is therefore to be expected that the lipid content of the diet will modify the composition and function of the endoplasmic reticulum of the liver containing the membrane-bound MFO system.

Indeed, feeding rats a lipid-free diet for three weeks has been shown to depress hepatic cytochrome P-450 concentrations and the microsomal drug-metabolising enzymes, compared with animals receiving a 3% corn oil diet (Norred and Wade, 1972). However, it is increasingly being realised that the quantity of lipid is of high importance in the regulation of the rate of biotransformation of foreign compounds. Thus, Hietanen and his associates (1975) have demonstrated that addition of 5–10% of olive oil to the diet of rats increases the activity of the MFO system, and 25–35% of dietary fat depresses the hepatic oxygenases, independent of the degree of saturation of fatty acids. In rats, a low dietary intake of corn oil (<1% of calories as linoleic acid) decreases the hepatic activity of the MFO system, and a high intake (9% calories as linoleic acid) is equally deleterious (Caster *et al.*, 1970). A similar effect is observed following intraperitoneal administration of the unsaturated fatty acid (40 mg/kg) to rats (Lang, 1976). The optimum dietary intake of linoleic acid for maximum activity of the MFO system appears to be about 3% of total calories.

Recent studies have indicated that it is not only the quantity but also the quality of lipid that is important in relation to the maximum activity of the MFO system. Thus, there appears to be no difference in the hydroxylation of benzo(a)pyrine between a fat-free and a 10% lard diet, whereas a 10% corn-oil diet increases it to about 35% (Lambert and Wills, 1977). These workers have also shown that a 5% lard diet, supplemented with 5% methyl linoleate, enhances the metabolism of benzo-(a)pyrine, suggesting the important role of the fatty acid in the diet in increasing the oxidative metabolism of foreign compounds. Furthermore the rate of induction of the hepatic MFO system by phenobarbital appears to be enhanced in rats whose diets are supplemented with unsaturated fatty acids (corn oil), compared with animals receiving saturated fatty acids (coconut oil) (Marshall and McLean, 1971).

The MFO system of the hepatic endoplasmic reticulum is concerned not only with the oxidative catabolism of drugs and environmental chemicals, but also with the oxidative catabolism of cholesterol to bile acids. High-cholesterol diets have been claimed to increase the incidence of chemical carcinogenesis of the colon and liver as a consequence of altered metabolism of the carcinogens and bile acids or changes in the gut flora (Rogers and Newberne, 1975). It is therefore important that an adequate dietary intake of unsaturated fatty acids is maintained to optimise the catabolism of cholesterol as well as the potentially toxic environmental chemicals.

2.2.3 Dietary carbohydrates

The depressed metabolism of various drugs in man as a consequence of high carbohydrate with less protein intake has been discussed earlier. In recent years there have been a number of experimental reports showing that, in addition to the quantity the quality of carbohydrate may be important in relation to its modifying effect on the biotransformation of foreign compounds. Thus, feeding sucrose instead of starch markedly decreases in rats the activities of aminopyrine N-demethylase, pentobarbital oxidase, aniline and biphenyl-hydroxylases, and neoprontosil azo-reductase in parallel with hepatic concentrations of cytochromes P-450 and b_5, NADPH oxidase, and NADPH-cytochrome c reductase (Jarvell *et al.*, 1965; Kato, 1967; Basu *et al.*, 1975b). In accordance with the results *in vitro*, the metabolism of carisoprodol *in vivo* is decreased and strychnine mortality and carisoprodol paralysis are enhanced by sucrose feeding (Kato, 1967). Moreover, high-sucrose diets fed to rats have been reported to potentiate the lethal reaction of benzyl penicillin (Boyd *et al.*, 1970), perhaps due to reduced rates of conversion to its less toxic metabolite.

Since the consumption of sucrose is steadily rising (Yudkin, 1964), and drugs are often given to children in sucrose syrup, a realisation of the significance of the experimental findings in humans is certainly warranted. It is also important to recognise that the hepatic MFO system is concerned not only in the metabolism of foreign chemicals, but also in the metabolism of steroid hormones and cholesterol, the blood levels of which are known to be influenced by high sucrose consumption (Strother *et al.*, 1971).

A high intake of glucose, fructose or sucrose has also been reported to decrease hexobarbital hydroxylation and increase hexobarbital sleeping time in mice. On the basis of these results, it can be anticipated that intravenous administration of 5% glucose over an extended period of time to ill patients could have marked effects on barbiturate-induced sleeping time.

2.2.4 Deficiency of vitamins

There has been increasing experimental evidence suggesting that a variety of vitamins may also regulate the metabolic biotransformation of various foreign compounds. The vitamins which have been studied in relation to the hepatic microsomal MFO system are: riboflavin, ascorbic acid, retinol and α-tocopherol. Among these vitamins, ascorbic acid is perhaps the most extensively studied vitamin, and in species like man and guinea pig which have a dietary requirement for ascorbate the effect of deficiency on the metabolism of foreign compounds can be quite severe.

2.2.4.1 Ascorbic acid (ASA)

There is considerable evidence linking changes in the activity of the hepatic microsomal MFO system with changes in ASA status. These studies have revealed that guinea pigs deficient in the vitamin exhibit depressed levels of liver cytochrome P-450 and reduced capacities to metabolise drugs (Zannoni et al., 1972; Zannoni and Lynch, 1973; Rikans et al., 1978). Earlier studies have also indicated that scorbutic guinea pigs are more sensitive than normal guinea pigs to pentobarbital, procaine, and the muscle relaxant zoxazolamine (Conney et al., 1961).

Several attempts have been made in order to highlight the possible mechanism by which ASA deficiency affects the metabolism of drugs. Studies with typical enzymes such as N-demethylase and O-demethylase indicate that there is no significant change in the affinity of the enzymes for their substrates in normal, deficient or supplemented animals (Zannoni, 1977). Using inducers of the drug-metabolising enzymes, Rikans and associates (1978) have shown that de novo synthesis of protein is operable in the presence of ASA deficiency. In addition, Sato and Zannoni (1976) have shown that ASA-deficiency-mediated depressed drug metabolism is not due to an increase in lipid peroxidation or to a quantitative change in the essential phospholipid component in drug metabolism. However, these workers have observed that the differences that exist between normal and ASA-deficient microsomes include less stability of the ASA-deficient microsomes to sonication, dialysis and treatment with ferrous iron metal chelators such as α-dipyridyl and O-phenanthroline. Losses of cytochrome P-450 and O-demethylase activity during dialysis could be prevented by the addition of ASA, and the vitamin protects cytochrome P-450 against inhibition by ferrous iron chelators. These studies suggest that there may be an interaction between ASA and cytochrome P-450 involving haem iron. It is of further interest that restoration of normal levels of the drug-metabolising activities requires 6 days of ASA supplementation (Zannoni et al., 1972), whereas the cytochrome P-450 concentration is restored to control values within 48 h when ASA is supplemented with 6-aminolevulenic acid, the precursor of haem synthesis (Luft et al., 1972). These findings suggest that ASA participates in the biosynthesis of haem and that its effect on cytochrome P-450 concentration may be mediated through the haem cycle.

In a recent study, it has been demonstrated that changes in ASA intake markedly influence the rate of drug metabolism in man (Ginter and Vejmolova, 1981). This study involved nine healthy volunteers receiving 500 mg ASA per day and eight subjects receiving placebo, for one year. At the end of the period, following an overnight fast, all subjects were given 1 g antipyrine orally, and its biological half-life was determined. This study reveals a significantly higher value for the elimination constant and a shortened half-life of the drug in the ASA-treated group. These results indicate that ASA influences the ability to metabolise drugs not only in guinea pig but also in man.

ASA supplementation over an extended period has been shown to reduce the blood cholesterol level in hypercholesterolaemic subjects (Hanck and Weiser, 1977). This may be the result of ASA-mediated raised activity of cholesterol 7α-hydroxylase, a hepatic microsomal oxygenase which catalyses the oxidation of cholesterol to bile acids (Ginter, 1977). ASA may thus play a significant and fundamental role in atherosclerosis (Turley et al., 1976), in addition to its ability to detoxify foreign chemicals.

In recent years substantial evidence has been presented suggesting that a variety of experimental tumours of the alimentary tract, liver, lung and bladder can be produced by nitroso compounds (Mirvish, 1971; Mirvish et al., 1975; Rustia, 1975; Narisawa et al., 1976), which are produced by the reaction of nitrites with secondary and tertiary amines, amides or ureas.

$$(R_1, R_2)-N-H + HNO_2 \rightleftharpoons (R_1, R_2-N-N = O + H_2O \quad (Eq. 2.1)$$

In 1972, Mirvish and his co-workers demonstrated that the nitrosation reaction can be prevented by adding ASA in vitro. More recently, an in vitro study has provided further evidence as to the spontaneous nitrite-lowering effect of the vitamin, and that such an effect correlates with the rate of oxidation of ASA (Basu et al., 1984).

Experiments with rats have shown that the vitamin also exerts a protective effect against hepatotoxicity following oral administration of sodium nitrite and aminopyrine (Kamm et al., 1973). Hepatotoxicity in this case is believed to be due to formation of dimethylnitrosamine by the reaction of nitrous acid with aminopyrine. Fiddler and his co-workers (1973) have

33

demonstrated that addition of ASA to the ingredients used to cure frankfurters greatly reduces the formation of dimethyl-nitrosamine. Furthermore, under simulated gastric conditions (37°C, pH 1.5), an ASA/nitrite molar ratio of 4:1 appears to provide 93% protection from *in vitro* nitrosation of methylurea (Table 2.2).

Table 2.2: Effect of ascorbic acid (ASA) on the formation of methylnitrosourea in potato incubated with nitrite and methyl-urea under simulated gastric conditions

ASA/Nitrite (molar ratio)	Nitrite (% decrease)	Methylnitrosourea (% inhibition)
0/1	0	0
1/1	13	37
2/1	36	74
4/1	43	93

Modified data from Raineri and Weisburger (1975). Five-gram samples of homogenised boiled potato were used.

ASA does not seem to react with amines, nor does it increase the rate of nitrosamine decomposition. However, it reacts very rapidly with nitrite (Fiddler *et al.*, 1973). It is possible that the vitamin lowers the amount of available nitrite by reducing it to nitrogen oxide, and consequently inhibits the N-nitrosation reaction.

$$2HNO_2 + ASA \rightarrow Dehydro\ ASA + 2NO + 2H_2O \quad (Eq.\ 2.2).$$

Further evidence for the protective effect of ASA in carcino-genesis induced by nitrosamine formation has been provided by Guttemplan (1977). Using the Ames test, this worker has shown that the vitamin inhibits bacterial mutagenesis by N-methyl-N-nitroguanidine, the mutagenesis being a change which could potentially lead to carcinogenesis in the animal cell.

2.2.4.2 Other vitamins

Chronic riboflavin deficiency in rats has been shown to result in a significant decrease in drug-metabolising enzyme activities with type 2 substrates such as aniline and acetanilide and type 1 substrates such as aminopyrine and ethylmorphine (Patel and Pawan, 1974). The depressed activities of drug-metabolising enzymes also appear to parallel those of microsomal electron transport components including NADPH-cytochrome c reductase, cytochrome b_5, and cytochrome P-450 (Patel and Pawan, 1974; Yang, 1974; Hara and Taniguchi, 1982). It is of interest that it requires 10–15 days of riboflavin supplementation to restore the changes in the MFO system to normal (Patel and Pawan, 1973, 1974), suggesting that the effect on drug metabolism may be due to ultrastructural changes in the endoplasmic reticulum of the hepatocyte rather than a biochemical lesion (Tandler et al., 1968).

Retinol (vitamin A) deficiency in rats has also been shown to decrease the hepatic cytochrome P-450 along with the activities of drug-metabolising enzymes (Becking, 1973; Colby et al., 1975; Miranda et al., 1979). Retinol and its synthetic analogues are known to be potent agents for the control of cell differentiation in epithelial tissues.

Convincing evidence of the significance of retinol in protecting against cancer has come from experimental studies in which tissues in organ culture or in intact animals are exposed to carcinogenic polycyclic aromatic hydrocarbons to develop squamous cancer. In addition, a number of studies have shown that systemic prophylactic administration of the vitamin, both before and after exposure to various chemical carcinogens including 3-methylcholanthrene, benzopyrine and dimethylbenzanthracene, inhibits the induction of metaplasia and carcinoma in various sites (Chu and Malmgren, 1965; Saffiotti et al., 1967, Cone and Nettlesheim, 1973; Sporn et al., 1976). There is also evidence that in prostate culture the hyperplastic and anaplastic lesions induced by chemical carcinogens can be reversed by the addition of retinoids (Chopra and Wilkoff, 1975). In vitamin A deficiency the binding of benzopyrine to tracheal epithelial DNA appears to be markedly increased (Genta et al., 1974).

In α-tocopherol- (vitamin E) deficient rats and rabbits, there appears to be a marked decrease in the n-demethylation reactions, as in the metabolism of aminopyrine and codeine, and treatment with the vitamin appears to normalise the activities (Carpenter, 1972; Carpenter and Howard, 1974). However, only chronic tocopherol deficiency results in a significant reduction of microsomal cytochromes of rat liver (Hauswirth and Nair, 1975). Administration of the vitamin has also been shown to reduce the hepatotoxicity of paracetamol, whereas tocopherol deficiency potentiates the toxic effects of the drug (Walker *et al.*, 1974).

2.2.5 Deficiency of inorganic elements

Experimental evidence is now available to suggest that some minerals have roles in the metabolism of foreign compounds. Thus the dietary deficiencies of minerals, with particular reference to iron, calcium, zinc, magnesium, copper and potassium, have been shown to affect the rate of drug metabolism in animals. In view of the fact that clinical and subclinical deficiencies of these elements are increasingly being recognised in large segments of the world's population, the need for information on the effect of nutritional deficiencies on man's ability to metabolise drugs becomes evident.

2.2.5.1 *Iron*

Cytochrome P-450, being a haem-containing protein, requires iron for its biosynthesis. However, the microsomal MFO activity of the gastrointestinal tract appears to be more sensitive to iron deficiency than that of the liver. Thus, the hepatic concentrations of cytochromes P-450 and b_5 have been shown to be unaffected in iron-deficient rats, as indicated by reduction in their plasma haemoglobin concentration from 14.8 to 7.7 g/100 ml following an iron-deficient diet for 40 days (Catz *et al.*, 1970; Becking, 1972). In contrast, the concentration of cytochrome P-450 present in the villous tip cells of rat duodenal mucosa is reduced by 50% when the animals are fed an iron-deficient diet for only 2 days (Hoensch *et al.*, 1975).

It is noteworthy that iron has been reported to interfere with the absorption of tetracycline (Neuvonen *et al.*, 1970), suggesting the necessity to withhold iron supplementation when patients are receiving oral tetracycline preparations.

2.2.5.2 *Other inorganic elements*

Dietary deficiencies of calcium, magnesium, copper and zinc tend to affect one or all of the components of the MFO system (Dingell et al., 1966; Becking and Morrison, 1970a, b). However the sensitivity of the effect of these mineral deficiencies on drug metabolism depends upon the element. Thus, lowered rates of aminopyrine, hexobarbital and *p*-nitrobenzoic acid metabolism are observed in rats after 40 days on a calcium-deficient diet (Dingell *et al.*, 1966). In contrast, a magnesium-deficient diet reduces the rates of aniline and aminopyrine metabolism after only 11–14 days (Becking and Morrison, 1970b).

Another important intracellular mineral which plays a major role as a cationic electrolyte in the maintenance of membrane integrity is potassium (Becking, 1976). *In vivo* studies in rats have shown that hypokalaemia reduces significantly the plasma clearance of drugs, and increases the phenobarbital-induced sleeping time. The body content of potassium has also been reported to affect the toxicity of digitalis in patients with cardiac failure (Sodeman, 1965). Thus, any treatment (see Chapter 4) which reduces the blood potassium level in a digitalised patient will increase the risk of digitalis-induced cardiac arrhythmias (D'Arcy and Griffin, 1972). A loading dose of less than 1.0 mg digoxin may produce toxicity in the presence of hypokalaemia.

2.3 NON-NUTRITIONAL DIETARY FACTORS AFFECTING THE METABOLISM OF ANUTRIENTS

In addition to essential nutrients, food contains numerous non-nutritional components, such as nitroso compounds, food additives and environmental contaminants, including pesticide residues and polycyclic aromatic hydrocarbons (see Chapter 1). Many of these foreign compounds have been reported to have modifying effects on the intestinal and hepatic metabolism of drugs.

Benzopyrine is known to cause a marked increase in phenacetin metabolism in the gut (Kuntzman *et al.*, 1977). In contrast, benzopyrine and 3-methylcholanthrene appear to depress dimethylnitrosamine demethylase activity in the livers of rats. Pantuck and his colleagues (1976b) have demonstrated that rats maintained on a diet containing cruciferous vegetables such as sprouts and cabbage show a marked increase in the activities of intestinal enzymes involved in the metabolism of a variety of drugs. The vegetables contain various indole derivatives in large amounts, which are believed to be potent inducers of intestinal and hepatic drug metabolism.

The method of food preparation may also lead to an interaction with foreign compounds. Feeding a diet containing charcoal-broiled beef for 4 days has been shown to decrease plasma half-lives of antipyrine in man (Alvares *et al.*, 1976a, b). Further studies in man have demonstrated that a similar diet markedly depresses the plasma levels of orally administered phenacetin, but leaves the plasma half-life unchanged, implying that the diet stimulates the metabolism of phenacetin only in the gut (Conney *et al.*, 1976; Pantuck *et al.*, 1976a). Charcoal-broiled beef also enhances the intestinal metabolism of phenacetin and stimulates benzopyrine hydroxylase activity in the liver of pregnant rats (Harrison and West, 1971).

3

Nutritional Consequences of Drug Therapy

In recent years, there has been a growing awareness that treatments with drugs of widely differing chemistry and pharmacological action have the potential to cause nutritional deficiencies. The drug-induced deficiencies may range from overt depletion with clinical manifestation to sub-clinical deficiency, depending upon one's nutritional status at the onset of therapy, the type, dosage and duration of drug use, the presence of predisposing disease, and the age of the patient. The actual incidence of drug-induced nutrient deficiency is, indeed, difficult to identify, since cause-and-effect relationships between the offending drug administered and the nutrient deficiency are difficult to prove. However, there is evidence that certain specified drugs induce deficiencies of specific nutrients, and that these deficiencies occur following long-term, high-dose administration in susceptible patients. Many of these drugs may alter taste perception, reduce absorption, increase excretion or interfere with the utilisation of nutrients.

Some of these drug-nutrient interactions, because of their wide prevalence, cannot be ignored. However, the widespread consumption of certain drugs, often self-prescribed, makes it impossible to rely upon advice for each individual from a physician, even if the latter were aware of the interactions. It would therefore seem that for certain nutrients possible interaction with drugs should be considered in relation to needs on a community basis.

3.1 DRUGS AFFECTING FOOD INTAKE

There is a paucity of scientific studies of the influence of drugs on appetite in humans. A number of factors, emotional, psychological, pathological and economic, may all have effects on appetite and food intake and thus complicate the interpretation of experimental results. However, a variety of drugs are believed either to decrease or increase appetite through various mechanisms (Pawan, 1974).

Alcohol, when taken in a small amount before a meal, increases the appetite by stimulating the sense of taste and by increasing the flow of saliva and gastric and pancreatic secretions. Excessive intake of alcohol, however, appears to have the opposite effect, as is frequently seen among alcoholics. Insulin-induced hypoglycaemia is associated with increased appetite. Other hormones, such as androgens, anabolic steroids, glucocorticoids and thyroid hormones, are also known to increase appetite. In addition, the oral antidiabetic agents, especially sulphonylureas (such as tolbutamide, chlorpromamide and tolazamide), have been reported to increase appetite, possibly through their stimulating effects on insulin release (Pierpaoli, 1972; Yosselson, 1976). Several other drugs have been recognised as having appetite-promoting effects. In depressed subjects, an improvement in appetite is often seen following prolonged therapy with a high dosage of psychotropic drugs. The antihistamines, such as buclizine and cyproheptadine, have also been reported to cause an increase in food intake followed by weight gain in underweight adults (Drash *et al.*, 1966; Bergen, 1974).

A number of drugs have been recognised as affecting nutritional status by decreasing the appetite. Bulk agents, such as methylcellulose and guar gum, appear to have a moderate effect on reducing the appetite. They take up fluid and swell in the stomach, giving a feeling of fullness. Amphetamines and related drugs have been widely used as effective anorectic agents. The effect is believed to be mediated through catecholamine- or serotonin-dependent mechanisms (Datey *et al.*, 1973; Blundell *et al.*, 1976). Treatment of hyperactive children with amphetamines has been reported to be associated with growth retardation (Safer and Allen, 1973). The oral antidiabetic agents, particularly biguanides, may also reduce appetite and food intake (Malcolm *et al.*, 1972).

The anorectic effect of these drugs may be useful in the therapy of obese patients, but the clinician and dietitian must be aware of potentially harmful derangements of normal physiology. Thus, there are many reports suggesting that most of the anorexigenic drugs with particular reference to amphetamines have been extensively abused (Isbell and Chrusciel, 1970), because they produce an elevation of mood, a reduction of fatigue and a sense of increased alertness. The chronic and extensive use of these drugs may result in addiction and the development of unwanted nutritional derangements such as nausea, diarrhoea, vomiting and abdominal pain. Furthermore, like many antidepressant drugs (tricyclics), amphetamine (a synthetic sympathomimetic amine) is also a monoamine oxidase (MAO) inhibitor. The use of this drug in depressed patients who are already being treated with antidepressants may therefore precipitate severe side effects. For these patients the intake of tyramine-containing foods (Figure 3.1) may exacerbate even further the side effects of the anorexigenic amine (Natoff, 1965). Tyramine is a stimulant of the sympathetic

Figure 3.1: Food containing tyramine, and drug incompatibilities

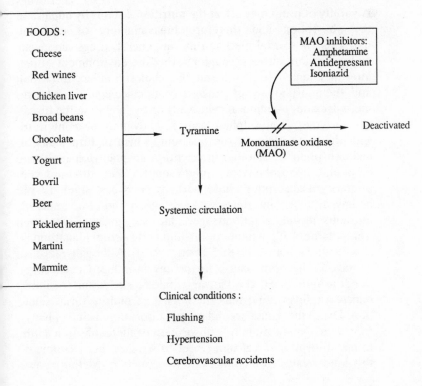

system and thereby it may cause a marked rise in blood pressure. The ingested tyramine is normally inactivated in the liver by MAO through oxidative deamination. However, when the enzyme is blocked by amphetamine or other drugs, the free tyramine can gain access to the circulation in large amounts and cause a clinical state characterised by diarrhoea, headache and hypertension. Amphetamine should therefore be used with caution in patients receiving antidepressants, and the tyramine-containing foods must be avoided by patients while they are being treated with the drugs.

In addition to the drugs mentioned above, several other drugs have been found to reduce appetite and food intake. These include chlorambucil (anticancer), diethylcarbamazin (antifilariasis), diethylstilboestrol (post-coital contraceptive), thiabendazole (antihelminthic) and glucagon (hyperglycaemic).

3.2 DRUGS AFFECTING ABSORPTION

A variety of drugs may affect the nutritional status by impairing the absorption and therefore bioavailability of essential nutrients via several mechanisms. In general, drugs can cause malabsorption, either through altering the environment of the intestinal lumen (e.g. mineral oil, cholestyramine, antacids) and the morphology of mucosa (e.g. colchicine, neomycin, methotrexate, p-aminosalicylic acid) or by inhibiting the digestive enzymes (e.g. sulphasolazine). They may also interfere with the metabolism of one nutrient, which in turn leads to malabsorption of another nutrient (e.g. phenytoin, phenobarbital, diphosphonates, prednisone). The drug-induced primary malabsorptions have perhaps been best studied with neomycin, which interferes with the absorption of a variety of nutrients including fat, nitrogen, lactose, glucose, carotene, iron, vitamin B12, sodium, potassium and calcium (Jacobson et al., 1960; Dobbins, 1968; Faloon, 1970). Although mucosal damage is the prime cause for neomycin-induced malabsorption, the antibiotic is also known to bind bile acids and to inhibit pancreatic lipase, and thus there exists a multifactorial situation. Drugs that cause secondary malabsorption include phenytoin, which alters vitamin D absorption or metabolism, leading to malabsorption of calcium (Roe, 1974). Another example of this kind relates to methotrexate and calcium. Methotrexate, which functions as an antifolate drug, impairs intestinal absorption of calcium (Craft *et al.*, 1977). A large variety of drugs are known to cause both primary and secondary malabsorption of various nutrients; this will be discussed in more detail in subsequent sections of this chapter.

3.3 DRUGS AFFECTING CARBOHYDRATE METABOLISM

Many drugs are known to have a wide variety of actions in animal tissues. Some of the most important are their effects on carbohydrate metabolism. Broadly speaking, these drugs can be divided into two groups:

(a) hypoglycaemic, which lower blood sugar level with no significant change in glucose tolerance, and
(b) hyperglycaemic, which raise blood sugar concentration with reduction in glucose tolerance.

3.3.1 Hypoglycaemic agents

Sulphonylureas and biguanides are the most potent agents in lowering blood sugar level. The action of these drugs on carbohydrate metabolism is exerted through stimulating the pancreas's own ability to secrete insulin. Hence, they are only effective in the maturity-onset type of diabetes when there is still some islet capacity left. As with insulin, overdosage with the oral antidiabetic agents can cause serious or fatal hypoglycaemia. The risks are even greater with chlorpropamide (one of the sulphonylureas), because of its longer half-life (Sackner and Balian, 1960). Elderly and malnourished individuals are particularly vulnerable to the adverse effect of sulphonylureas as their detoxification mechanisms and renal functions are impaired (Davidson, 1971). Furthermore, a number of drugs are known to compete with sulphonylureas for their binding sites on plasma protein (e.g. phenylbutazone, salicylic acid and warfarin) and for their metabolising enzymes in liver (e.g. sulphaphenazol and phenyramidole). As a consequence of the drug interactions, the hypoglycaemic effect of sulphonylureas could be further potentiated through increased half-life and decreased metabolic degradation of these drugs. This may account for a large proportion of the fatal accidents attributable to sulphonylurea-induced hypoglycaemia (Marks, 1974). In view of the possible hazards as a consequence of the treatment with oral hypoglycaemic agents in patients with diabetes mellitus, it has been considered to be unwise to use any of the agents unless diet therapy is optimal and insulin therapy is impossible (Allan, 1971).

In addition to diabetes mellitus, Parkinson's disease and multiple sclerosis may also be treated with sulphonylureas, especially tolbutamide. The latter's use in an elderly non-diabetic patient with Parkinson's disease has been reported to result in severe and prolonged hypoglycaemia, leading to coma and focal neurological signs (Schwartz, 1961).

Other agents which have been found to lower blood glucose level include propanolol, phenylbutazone, barbiturates, alcohol and aspirin. The ability of the last two agents to produce severe and occasionally fatal hypoglycaemia should be given greater importance in view of the fact that the agents are readily accessible to the public.

Alcohol-induced hypoglycaemia is thought to be mediated through inhibiting hepatic gluconeogenesis (Field *et al.*, 1963); it is often seen when gluconeogenesis is the major source of energy entering the blood. Consequently, normal individuals who have fasted for periods of 2–3 days (Field *et al.*, 1963; Freinkel *et al.*, 1963), patients with thyrotoxicosis (Arky and Freinkel, 1966), diabetics with uncontrolled hyperglycaemia and ketonaemia (Arky and Freinkel, 1964) and individuals with diminished adrenocortical function (Arky and Freinkel, 1966), are particularly vulnerable to the hypoglycaemic action of alcohol. Chronic administrations of acetylsalicylic acid, the principal component of aspirin, have also been found to have a profound effect in lowering blood glucose level in diabetic and fasted normal subjects (Hecht and Goldner, 1959). Like alcohol, the mechanism of action of salicylate is mediated through inhibiting gluconeogenic processes. Experimental studies have shown that salicylates decrease hepatic concentrations of glycogen and the key regulatory enzymes involved in gluconeogenesis–pyruvate carboxylase and phosphoenol pyruvate carboxylinase (Madappally *et al.*, 1972).

In addition to the drugs mentioned above, here are a number of other drugs commonly used in medical practice which may also lower blood glucose level by affecting the digestion and absorption of various carbohydrates (Gray, 1973). These include neomycin, colchicine, digitoxin and elhacrynic acid).

3.3.2 Hyperglycaemic agents

Hyperglycaemia results from the insulin secreted by the pancreas being either insufficient in amount or inefficient in action for one or more reasons. A large variety of drugs are known to have hyperglycaemic properties. These include corticosteroids, oral contraceptives and thiazide diuretics. An effect of corticosteroids on carbohydrate metabolism in humans is well recognised because of the frequency of impaired glucose tolerance in Cushing's syndrome and of hypoglycaemia in adrenal insufficiency. It seems that all corticosteroids influence carbohydrate metabolism, leading to hyperglycaemia (Matsunage et al., 1963; Eisenstein, 1973). The magnitude of action, however, is determined by the structural characteristics of the hormones. Steroid molecules with hydroxyl groups in the 17-α and 11-β positions appear to possess the most potent diabetogenic action. The mechanism of action of the corticosteroids on carbohydrate metabolism is believed to be exerted in several ways. The hormones are involved in promoting gluconeogenesis (Eisenstein et al., 1966) and glycogen synthesis (DeWulf et al., 1970), and possibly in inhibiting glucose utilisation (LeCocq et al., 1964).

Relative impairment of oral and intravenous glucose tolerance in women taking oral contraceptives has been recognised in many cross-sectional and longitudinal studies (Gershberg et al., 1964; Buchler and Warren, 1966; Wynn and Doar, 1969). In some cases, the changes in carbohydrate metabolism are believed to be similar to those seen in the later stages of pregnancy (Wynn and Doar, 1966). The increased blood glucose level, however, has been found to be restored to normal when pregnancy is terminated (O'Sullivan and Mahan, 1964) or when oral contraceptives are discontinued (Wynn and Doar, 1966, 1969). Nevertheless, the risk of precipitating permanent diabetes by the use of oral contraceptives on a long-term basis cannot be ignored on the basis of these studies. Prolonged longitudinal studies are certainly needed to determine whether oral-contraceptive-treated subjects with sub-clinical diabetes may later become permanently diabetic.

Indeed, the incidence of impaired glucose tolerance in women receiving oral contraceptives is thought to be highest in those with a positive family history of diabetes (Wynn and Doar, 1966). This claim has been supported by a study of the effect of a combination-type oral contraceptive on the carbohydrate metabolism of women who had abnormal glucose tolerance during pregnancy but normal glucose tolerance afterwards (Szabo et al., 1970). In this study, five women received a combination of norethindrone and mestranol as an oral contraceptive agent. In all five women, abnormal glucose tolerance recurred and three of them remained diabetic even when contraceptive therapy ceased. Similar observations were made with corticosteroid therapy (Seltzer, 1970). Thus, 26% of close relatives of diabetic individuals whose glucose tolerance was impaired by administration of cortisone developed frank diabetes within 15 years, but only 3% of close relatives whose glucose tolerance was not affected by cortisone became diabetic.

It appears that pregnancy-, oral-contraceptive- and corticoid-induced abnormalities of carbohydrate metabolism are manifest in humans who have inherited a trait that predisposes them to become diabetic. It may be that these subjects have an impaired capacity for insulin secretion, and they may also be abnormally resistant to the action of insulin. On the basis of these results, it may not be unreasonable to suggest that screening tests should be performed for diabetes, perhaps in the third trimester of pregnancy or early during contraceptive therapy, as an attempt to identify women with latent diabetes and exclude them from hormonal birth control (Szabo et al., 1970).

Prolonged and extensive use of thiazide diuretics (Table 3.1) has been found to be associated with hyperglycaemia (Lambert, 1975). A study in humans (Fajans et al., 1966) has indicated that trichlormethiazide-induced hyperglycaemia may be due to a

Table 3.1: Thiazide diuretics with proprietary name

Chlorothiazide (Diuril)
Hydrochlorothiazide (Esidrix)
Hydroflumethiazide (Naturetin)
Methylclothiazide (Enduron)
Benzthiazide (Exna)
Cyclothiazide (Anydron)
Polythiazide (Renese)
Trichlormethiazide (Metahydrin)

decrease in the amount of endogenous insulin release. There is an impairment of the insulin effect on both muscles (Barnett and Whitney, 1966) and adipose tissue (Field and Mandell, 1964). Hypokalaemia, induced by thiazide diuretics (see later), is also believed to be a contributory factor for resulting hyperglycaemia.

The hyperglycaemic effect of thiazides may therefore result in the loss of diabetic control in patients being treated with oral hypoglycaemic agents or insulin or whose condition is being controlled with the diet alone. It appears that approximately 10% of patients treated with oral hypoglycaemic agents require either an increase in dosage or a change to insulin treatment while being treated by thiazide diuretics (Kansel *et al.*, 1969).

3.4 DRUGS AFFECTING LIPID METABOLISM

Drugs that affect lipid metabolism can be divided into two groups: (a) hypolipidaemic agents which lower blood lipid levels, and (b) hyperlipidaemic agents which raise blood lipid concentrations. Drug therapy in hyperlipidaemia is applied for those patients whose dietary response is inadequate. The hyperlipidaemic actions of certain drugs are normally regarded as undesirable side effects, which may or may not be predictable.

3.4.1 Hypolipidaemic agents

A wide variety of drugs are now available to lower blood lipid levels in patients with hyperlipidaemia. These drugs are effective in hypercholesterolaemia or in hypertriglyceridaemia or in both.

3.4.1.1 Hypocholesterolaemic drugs

The drugs that are used for the treatment of hypercholesterolaemia include cholestyramine, colestipol, β-sitosterol, *D*-thyroxine, *p*-aminosalicylic acid and neomycin. Cholestyramine and neomycin are the two drugs that are most commonly used in medical practice.

Cholestyramine is an anionic polymeric resin, which binds bile acids in the bowel and promotes their faecal excretion

(Hashim and Van Itallie, 1965; Moore *et al.*, 1968). This is followed by enhanced hepatic cholesterol and bile acid synthesis (Havel and Kane, 1973). The higher rate of elimination of sterols from the liver to the faeces via bile is associated with increased rates of synthesis of very low density lipoproteins (VLDL) and breakdown of low density lipoproteins (LDL) (Langer *et al.*, 1969). Hence, resin appears to lower the plasma cholesterol concentration and consequently its principal constituents, LDL, by up to 30%. Cholestyramine is a nonsystemic drug. Although it is known to be relatively non-toxic, its therapeutic doses of 8 to 16 g/day can induce steatorrhoea (Hashim *et al.*, 1961) and thereby may affect the absorption of various nutrients, especially the fat-soluble vitamins. Treatment with this drug on a long-term basis may also lead to iron-deficiency anaemia (see later).

Like cholestyramine, neomycin (a polyene antibiotic) when ingested orally (the dose being 1 to 2 g/day) lowers the cholesterol level by binding and promoting faecal excretion of its degraded products (Samuel *et al.*, 1967). The hypocholesterolaemic effects may be due to the development of cholesterol malabsorption, by interference with either micellar solubilisation or mucosal transport (Sedaghat *et al.*, 1975). However, neomycin, when taken by mouth, not only causes malabsorption of lipid and lipid-soluble substances, but also produces abdominal discomfort. *D*-Thyroxine, the dextro-isomer of the natural thyroid hormone, is also known to be an effective LDL–cholesterol lowering agent (Strisower and Strisower, 1964) by increasing cholesterol turnover and lipoprotein synthesis and removal (Kritchevsky, 1960). Doses of 4 to 6 mg/day lower plasma LDL by about 20%. Use of the drug in higher doses has been found to be often associated with high metabolic rate and even frank hyperthyroidism (Eisalo *et al.*, 1963; Jepson, 1963). It should, however, be pointed out that treatment of patients with myocardial infarctions with *D*-thyroxine has been reported to be associated with death rates higher than those occurring in patients treated with either placebo or other medications (Coronary Drug Project Research Group, 1972). *D*-Thyroxine is therefore contraindicated in patients with known heart disease.

3.4.1.2 *Hypotriglyceridaemic drugs*

Clofibrate (ethyl chlorophenoxyisobutyrate or Atromid S) is

probably the most widely used drug for the treatment of hypertriglyceridaemia. It lowers VLDL-rich triglycerides (up to 40%) as well as LDL-rich cholesterol (up to 20%) in plasma. Its mode of action is believed to be mediated, at least in part, through reduction in the rate of hepatic secretion of newly synthesised lipoproteins and inhibition of cholesterol synthesis (Azarnoff *et al.*, 1965; Nestel *et al.*, 1965). Clofibrate may also lower the plasma levels of free fatty acids (Macmillan *et al.*, 1965) by decreasing their synthesis (Maragoudakis, 1970), and this could account for the decreased VLDL level since free fatty acid is an important precursor of lipoprotein (Nestel and Whyte, 1968).

Clofibrate is a systemic drug, and its therapeutic dose is 2 to 3 g/day. It is relatively non-toxic. The only nutrition-related side effect reported in some patients is nausea, which may, however, disappear with continued treatment.

Nicotinic acid in large doses (3 to 6 g/day) is a potent depressant of plasma triglyceride levels. The mode of action of this vitamin is not known. However, it has been suggested that nicotinic acid inhibits the release of VLDL by reducing free fatty acid turnover (Carlson *et al.*, 1968; Bencze, 1975). Also, it seems to have a cholesterol-lowering effect, presumably by reducing the synthesis of cholesterol (Miettinen, 1968) and the conversion of VLDL to LDL (Carlson *et al.*, 1968). The use of large doses of nicotinic acid may be associated with potentially serious side effects including flushing, itching, impaired liver function, diarrhoea, raised serum uric acid, and hyperglycaemia (see Chapter 5).

3.5 DRUGS AFFECTING PROTEIN AND AMINO ACID METABOLISM

A number of drugs are known to affect the metabolism of proteins and amino acids (Table 3.2). Only a few appear to stimulate protein synthesis, but the majority have the opposite effect.

Glucocorticoids appear to cause a depletion of muscle protein and a deposition of protein in the liver (Munro, 1964), and the activity of the amino acid activating enzyme is increased in the liver of rats given a low-protein diet (Stephen, 1968) and in the liver of malnourished children (Waterlow, 1968).

49

Table 3.2: Drugs affecting protein metabolism

Drugs	Action
Corticosteroids	Increase gluconeogenesis and urinary nitrogen excretion
Chloramphenicol	Inhibits amino acid transfer during protein synthesis
Tetracyclines	Increase urinary nitrogen excretion and blood urea nitrogen
Oral contraceptives	Impair vitamin B_6 status, altering plasma amino acid levels
Indomethacin	Decreases amino acid absorption
Neomycin	Decreases amino acid absorption
Insulin	Stimulates protein synthesis
Anabolic steroids	Stimulate protein synthesis
Thyroid hormones	Increases urinary nitrogen excretion
Salicylates	Produce amino aciduria

Increased blood cortisol levels have also been found to be associated with malnourished children (Alleyne and Young, 1967) and animals (Basu *et al.*, 1975a). In addition, the adrenal cortex has been suggested to play a central role in the evolution of protein-energy malnutrition (Jaya Rao, 1974).

The catabolic action of glucocorticoids on protein metabolism is also responsible for muscle wasting and weakness of patients with Cushing's syndrome. Despite clinical evidence of protein wasting, these patients remain in nitrogen balance if the diet contains 1 g protein/kg body weight. In patients with Cushing's syndrome a high-protein diet results in positive nitrogen balance, whereas a low-protein diet causes negative balance.

Chloramphenicol, because of its structural similarity to phenylalanine (an essential amino acid), may compete with it, leading to an inhibition of protein synthesis (Carter and McCarty, 1966). The antibiotic, along with other drugs including penicillin and sulphonamides, may also bind with serum albumin, and thereby interfere with the normal carrier function of serum proteins (Goldstein, 1949).

Tetracycline is another antibiotic which has been found to have a profound effect on protein metabolism. Thus an analysis of data of 1957 patients revealed a strong association between tetracycline administration and development of clinically significant rises in blood urea nitrogen level (Boston Collaborative Drug Surveillance Program, 1972). However, the association was found to be confined to tetracycline recipients who also received diuretics.

An association with a significant fall in plasma concentrations of a variety of amino acids including alanine, glycine, leucine, methionine, phenylalanine, proline, tyrosine and valine has been reported in women taking oral contraceptives (Aly *et al.*, 1971; Craft and Peters, 1971). Since these subjects had a normal aminoaciduria, it has been suggested that oral contraceptives may be involved in increasing the utilisation of amino acids. The mechanism of action may be mediated through the oral contraceptive steroid hormone-induced vitamin B6 deficiency (see later).

3.6 DRUG AND VITAMIN INTERACTIONS

It may seem strange to suggest that drugs should enter into a consideration of the physiological requirements for vitamins. However, lack of adequate dietary vitamin intake is not the only way in which deficiencies occur. A number of secondary mechanisms exist, not the least of which is drug-induced vitamin deficiency. Administration of drugs producing 'relative' deficiencies in otherwise healthy populations may not be much of a problem, but administration of these drugs in patients with superimposed disease processes or in subjects with impaired nutritional status can result in hypovitaminosis with symptoms of avitaminosis. In view of the ready availability of drugs prescribed by clinicians, combined with the ease with which many pharmaceutical preparations can be obtained over the counter without a prescription, the drug-induced vitamin deficiencies deserve more attention than they have so far received.

Interaction between vitamins and drugs is influenced by the effects of drugs on the functions of vitamins and on the enzymes which they control. Drugs may also affect bacterial synthesis in the gastrointestinal tract as well as the rates of absorption, utilisation and elimination of vitamins.

3.6.1 Ascorbic acid

Several drugs, including salicylates, oral contraceptives, tetracycline, corticosteroids and calcitonin, have been reported to interact with ascorbic acid. Among these drugs, salicylate-containing agents such as aspirin are perhaps the most commonly and chronically used drugs, and hence impose a serious problem in the community at large. The interactions between aspirin and ascorbic acid have been the subject of many studies. Thus, administration of therapeutic doses of aspirin to healthy volunteers has been found to decrease the metabolic availability of ascorbate. Isolated reports have indicated that rheumatoid arthritic patients ingesting high doses of aspirin have low levels of plasma ascorbic acid (Basu, 1981).

In order to highlight the possible mechanism for the ascorbate–aspirin interaction, Basu (1981) investigated the effect of soluble aspirin on utilisable ascorbic acid in human subjects. In this study, the concentrations of ascorbate in plasma, leucocytes and urine were found to be markedly elevated at various intervals following administration of a single oral dose of 500 mg of the vitamin. The ascorbate-associated increases, however, appeared to be blocked when the vitamin was given simultaneously with aspirin (900 mg). Similar findings were observed in guinea pigs, where, in addition, faecal excretion of vitamin C was found to be significantly increased when the vitamin was administered orally together with aspirin.

The respiratory tract is a major route for elimination of vitamin C in guinea pigs. About 60–70% of the vitamin is believed to be metabolised to CO_2 and expired in the breath of these animals. Using radiolabelled vitamin C, the rate of exhalation of CO_2 has been shown to be three times higher in animals receiving vitamin C alone than in animals given the vitamin simultaneously with sodium salicylate (Ioannides *et al.*, 1982). These findings tend to indicate that the metabolic availability of vitamin C may be reduced in the presence of aspirin, possibly by impeding gastrointestinal absorption. Indeed, using gut sac preparations from rats, it has been shown that both the concentration gradient and the tissue uptake of vitamin C are markedly depressed in the presence of acetyl salicylate (Basu, 1981). Furthermore, using a segmental perfusion technique, it has been shown that jejunal glucose and sodium absorption rates are markedly inhibited after oral administration of 2–6 g aspirin to six healthy volunteers (Arvanitakis *et al.*, 1977). The aspirin-mediated decreased absorption rates of these substances have further been found to be accompanied by a significant drop in mucosal ATP levels measured in jejunal mucosal biopsies, indicating that aspirin may competitively inhibit the Na^+-dependent active transport of vitamin C.

These human and experimental findings tend to indicate that chronic ingestion of aspirin could be a serious problem in individuals with a borderline intake of vitamin C. There are many conditions, such as rheumatoid arthritis, for which the analgesic agent is most frequently used for a prolonged period. These conditions are often associated with an increased requirement and lowered blood levels of vitamin C (Basu, 1981).

There is evidence that many steroidal agents may also cause vitamin C deficiency. In 1972, Bartholomew reported a case of scurvy apparently precipitated by 8 days' prednisone treatment in a patient with rheumatoid arthritis. This patient was admitted with worse symptoms after 3 months' treatment with beta-methazone and indomethacin. Indeed, the clinical features of rheumatoid arthritis have been linked to those of scurvy, and it could be speculated that an apparent deterioration may have been due to progressive vitamin C deficiency aggravated by drug therapy.

Several studies have shown that there is an association in women receiving steroid-containing contraceptives with reduced levels of vitamin C in plasma, leucocytes and thrombo-cytes (Briggs and Briggs, 1973; McLeroy and Schendel, 1973; Rivers, 1975). In a recent study involving guinea pigs (Basu, 1986), oral administration of either oestrogen (5 μg) or progestogen (250 μg) in combination with 5 mg vitamin C (minimum requirement) daily for 21 days resulted in significantly lower concentrations of vitamin C in plasma, liver, adrenals and urine than in animals receiving only 5 mg of the vitamin. Furthermore, clinical manifestations of scurvy were exhibited in animals receiving no vitamin C but receiving the steroid hormones for 7 days. On the other hand animals receiving neither vitamin C nor the steroids remained free from any signs of scurvy. An *in vivo* dose-related effect of vitamin C indicated that the steroid-mediated lowering effect of the vitamin status could be counteracted by increasing the dose of vitamin C from 5 to 10 mg/day (Basu, 1986). It is evident, therefore, that interactions between oral contraceptives and vitamin C are of clinical importance, especially in women with inadequate or borderline intake of the vitamin.

The mechanism for the interaction is not yet known. However, serum levels of copper (Clemetson, 1968; Elgee, 1970) and ceruloplasmin (Carruthers *et al.*, 1966) have been shown to be elevated in these subjects. Since ceruloplasmin, the copper-containing protein, has ascorbate oxidase activity (Osaki *et al.*, 1964), the reduced vitamin C status may be due to increased breakdown of the vitamin. The steroid-associated biochemical evidence of vitamin C deficiency may also be due to decreased absorption of the vitamin (McLeroy and Schendel, 1973) and its altered distribution in tissues (Briggs and Briggs, 1973), or due to a decreased concentration of reducing substances such as glutathione (Saroja *et al.*, 1971). None the less, most recent studies have provided evidence that oral contraceptive steroids, especially oestrogen, may be responsible for decreasing absorption (Basu *et al.*, 1986) as well as increasing oxidation (Basu, 1986) of vitamin C.

There is evidence that tetracycline therapy over a 4-day period significantly reduces the leucocyte concentration but increases the urinary excretion of vitamin C, reflecting depletion of tissues (Shah *et al.*, 1968; Windsor *et al.*, 1972). It is possible that long-term therapy with this antibiotic could cause serious vitamin C depletion.

Calcitonin is another agent which has been found to have an effect on vitamin C metabolism. Thus, in a study involving five hospital patients with Paget's disease of the bones, it was apparent that intramuscular administration of porcine calcitonin (160 MRC units/day) for 3 weeks resulted in a decreased level of hydroxyproline excretion. This decrease was accompanied by a parallel drop in vitamin C excretion (Basu *et al.*, 1978). As far as could be determined, all patients were having a similar diet throughout the experimental period, and therefore it is unlikely that the decrease in vitamin C excretion with calcitonin could be associated with changes in the diet.

It was of further interest that the extent of pain relief in patients with Paget's disease was found to be markedly greater when calcitonin and vitamin C (3 g/day) were used together than when using calcitonin alone at 160 units/day (Basu *et al.*, 1978), indicating that calcitonin and vitamin C may be linked at some point in their actions.

There appears to be a reduced metabolic availability of vitamin C among cigarette smokers (Pelletier, 1975; Hoefel, 1977). A number of components are present in tobacco smoke which might be involved in the reduction of vitamin C bio-availability, namely nicotine, carbon monoxide, acetaldehyde, NO_2 and others. Among these substances, nicotine is the main one that induces cigarette smokers to smoke. Nicotine-induced vitamin C oxidation has been shown in *in vitro* studies (Pelletier, 1968). The biosynthesis of serotonin from *L*-tryptophan entails a hydroxylation step in which vitamin C may participate while being oxidised to dehydroascorbate. It is possible that there could be nicotine-induced irreversible oxidation of some of the dehydroascorbate (i.e. diketoglutonic acid) formed during the smoking-mediated biosynthesis of serotonin.

At steady state concentrations, smokers appear to have a higher metabolic turnover of vitamin C than do non-smokers (Kallner *et al.*, 1981). The turnover of the vitamin in smokers is approximately 90 mg/day, which is 50% higher than in non-smokers. The half-life of vitamin C has been shown to be shorter in smokers than in non-smokers, and the ratio between the amount of labelled vitamin C and total radioactivity is higher in non-smokers than in smokers (Kallner, 1981). These results provide further support for the high degree of metabolism of the vitamin among cigarette smokers.

3.6.2 Vitamin B6

Vitamin B6 functions as a coenzyme in the decarboxylation, transamination and deamination of most amino acids. A variety of drugs, notably isoniazid, hydrallazine and penicillamine, have been found to be antagonistic to vitamin B6 as a result of their structural similarity to the vitamin.

Hydrazides, such as the antitubercular drug isoniazid and the antihypertensive agent hydrallazine, form hydrazone complexes with the aldehyde form of vitamin B6 (pyridoxal). The formation of pyridoxal hydrazone results in enhanced urinary excretion and subsequent depletion of vitamin B6 from the body. Within a year of the introduction of isoniazid for treatment against tuberculosis, it was reported that patients treated with the drug developed peripheral neuropathy (Jones and Jones, 1953). Patients with impaired capacity to metabolise isoniazid to its inactive product, acetylisoniazid, are more prone to develop neuropathy than those who are rapid metabolisers

(Evans *et al.*, 1960). Penicillamine, an agent used to treat severe rheumatoid arthritis, Wilson's disease, cystinuria and lead intoxication, also forms a highly soluble complex with pyridoxal which is then rapidly excreted by the kidney. Vitamin B6 deficiency can result after prolonged administration or excessive doses of penicillamine (Jaffe, 1969).

The use of oral contraceptive agents (OCA) may be accompanied by increased urinary excretion of tryptophan metabolites, particularly after a tryptophan load (Aly *et al.*, 1971; Luhby *et al.*, 1971; Leklem *et al.*, 1975; Rose, 1978); increased *in vitro* stimulation of erythrocyte aspartic acid transaminase (ASPAT) (Aly *et al.*, 1971; Salked *et al.*, 1973; Rose *et al.*, 1973); depression (Herzberg *et al.*, 1970; Adams *et al.*, 1974; Rose, 1978); hypertriglyceridaemia (Rose *et al.*, 1977); impaired glucose tolerance (Spellacy *et al.*, 1972; Adams *et al.*, 1976; Rose, 1978); and a variety of symptoms of malaise (Winston, 1973). Plasma vitamin B6 concentrations fall in women using OCA, but this is thought to be only temporary (Lumeng *et al.*, 1974).

Whether the use of oral contraceptives produces a true vitamin B6 deficiency remains equivocal (Brown *et al.*, 1975; Leklem *et al.*, 1975). Nevertheless, 15–20% of OCA users show direct biochemical evidence of a vitamin deficiency (Rose, 1978). Barker and Bender (1980) state that whereas sub-clinical deficiency of vitamin B6 may be quite widespread, clinical deficiency is extremely rare. Recent studies using a depletion–repletion technique and a variety of biochemical indices indicate that the vitamin B6 requirement for most OCA users is approximately the same as that for non-users (Bosse and Donald, 1979; Donald and Bosse, 1979). The current evidence thus does not appear to justify the routine supplementation of dietary vitamin B6 with pyridoxine. According to Leklem *et al.* (1975), if the use of OCA does not alter the requirement for vitamin B6, the effect appears to be a minor one and of doubtful clinical significance for the majority of women ingesting these agents.

In 1980, the Foods and Nutrition Board of the National Research Council set the vitamin B6 recommended allowance for non-pregnant women at 2.0 mg. Some researchers suggest that this is adequate for OCA users with no other health problems affecting nutritional status (Brown *et al.*, 1975). Others propose that women who fail to adapt to OCA by not returning to normal blood B6 levels after 6 months of use, and those with low intake of vitamin B6, should use supplements (Thorp, 1980). Women who become pregnant after long-term use of OCA (3 years or more) may also be at risk, because post-partum maternal serum, cord serum and milk have been shown to exhibit a lower level of vitamin B6 than normal (Roepke and Kirksey, 1978).

General supplementation has not been advised because of the potential dangers associated with high intakes of vitamin B6 (Prasad et al., 1976); however, Faizy and co-workers (1975) propose a 10 mg supplement on the basis of the amount needed to prevent the lesion in the erythrocyte ASPAT test (Faizy and Mahatab, 1976). Kishi et al., (1977) found that only women taking 25–100 mg vitamin B6 daily obtained the 'ceiling of saturation' for erythrocyte ASPAT that is indicative of a state of no deficiency.

At present, vitamin B6 supplements seem to be advised only for high-risk groups in whom other factors, such as diet, pregnancy and disease, could increase the chances for a deficiency to develop.

It is obvious that more research is necessary in order to prove or disprove that a deficiency of vitamin B6 significant enough to be of concern to human health can actually be caused by OCA; to discover new biochemical parameters for assessing nutritional status more accurately; and to determine which groups of women may need supplements and in what amounts.

There is evidence that ingestion of vitamin B6 nullifies the beneficial effects of levodopa in the control of Parkinson's disease (Cotzias, 1969; Calne and Sandler, 1970). The parkinsonian syndrome is associated with histological lesions of the substantia nigra and with a depression of the dopamine content of the caudate nucleus (Hornykiewicz, 1966). Levodopa rather than dopamine is used because it crosses the blood–brain barrier more readily than dopamine (Gey and Pletscher, 1964). It is metabolised to dopamine by dopa decarboxylase, which requires pyridoxal phosphate as a coenzyme. This reaction takes place in the brain as well as in peripheral tissues. It is possible that, in the presence of vitamin B6, much of the levodopa is decarboxylated in extra-cerebral tissues and, as a result, a smaller amount of the drug enters the brain to replenish striatal dopamine. Current knowledge indicates that the peripheral decarboxylation of levodopa is accelerated by vitamin B6, probably in the gastrointestinal tract (Leon *et al.*, 1971; Mars, 1974). The degree of inhibition of the pharmacological effects of levodopa by vitamin B6 is dose related and can occur with as small a dosage as 5 mg. Patients receiving levodopa therapy are therefore recommended to avoid multiple-vitamin preparations and other preparations containing levels of vitamin B6 greater than 2 mg. It should, however, be pointed out that the concurrent use of carbidopa, a peripheral dopa decarboxylase inhibitor, prevents the levodopa-inhibiting effects of vitamin B6, and the carbidopa–levodopa combination product is recommended for use in patients receiving B6 supplementation.

3.6.3 Folic acid

In recent years, there has been a growing recognition of folacin deficiency induced by the use of anticonvulsants, as indicated by numerous publications describing the relationship between folacin deficiency and anticonvulsant therapy. This folacin deficiency and, in more severe cases, megaloblastic anaemia, have been related to phenytoin (Webster, 1954; Gydell, 1957), phenobarbital and primidone (Fuld and Moorhouse, 1956; Christensen *et al.*, 1957; Newman and Summon, 1957; Chanarin *et al.*, 1958) and to a combination of these drugs (Barker, 1957; Hawkins and Meynell, 1958; Druskin *et al.*, 1962). Although megaloblastic anaemia in epileptics taking anticonvulsants appears to be relatively rare, occurring in less than 0.75% of patients (Reynolds, 1973), the incidence of low serum folate levels in this group has been found to range from 27 to 91% in various studies (Reynolds, 1976).

That the observed megaloblastic anaemia in treated epileptics is directly related to the effect of anticonvulsant therapy on folate metabolism is supported by a number of investigators. Klipstein (1964) studied haematological changes in 65 epileptics, 60 of whom were receiving anticonvulsants. Of those who were taking diphenylhydrazine (DPH) and had serum folate levels less than 5.0 ng/ml, 74% showed some degree of macrocytosis, as assessed by peripheral blood films, and 18% who were taking DPH and had normal serum folate levels presented with macrocytosis. The macrocytosis could not be attributed to a deficiency of vitamin B12, since serum levels of this vitamin were found to be normal in all patients. It was of further interest that five untreated epileptics had normal serum and erythrocyte folate levels.

In more recent years the acute effect of phenytoin on serum folate concentrations was investigated by a number of workers. Thus, five out of six subjects consuming 1600 mg of DPH-sodium over a 4-day period showed a gradual fall in serum folate as measured by *Lactobacillus casei* assay (Richens and Waters, 1971). After cessation of drug intake, serum folate returned to near normal concentrations. The lowered serum folate was also shown to be significantly correlated with reduced red cell folate concentrations (Preece *et al.*, 1971). These results did not agree with the results obtained by Jensen and Olesen (1969), who found whole blood folate concentrations to be normal in treated epileptics. However, the lower range considered to be normal (20–100 ng/ml) in the latter study was substantially less than that (40–45 ng/ml) used by others (Hallstrom, 1969; Kallstrom and Nylof, 1969). Had Jensen and Olesen (1969) used 40 ng/ml as the lower limit for normal whole blood folate, two-thirds of the patients in their study would have been classified as being folate deficient.

In addition to haematological evidence, other biochemical studies have provided further support for the relationship between anticonvulsants and folate metabolism. Urinary excretion of formiminoglutamic acid (FIGLU), an intermediate in the catabolism of histidine, is an indicator of folate status. Normally, following an oral loading dose of histidine, urinary FIGLU excretion starts to increase in the second hour, reaches its peak after three to four hours, and gradually declines to base-line levels over the next few hours (Chanarin, 1980). Individuals with folate deficiency have an increased excretion of FIGLU, since the formiminotransferase-catalysed transfer of the formimino group to tetrahydrofolate is diminished as a result of low folacin levels. In a study conducted by Jensen and Olesen (1969), the mean level of urinary FIGLU excretion of 56 patients receiving DPH was 50% lower than in the control group. Ch'ien and colleagues (1975) also failed to show elevated FIGLU excretion after a test load of histidine in patients receiving DPH. A suggested explanation for these findings may be that DPH blocks the degradation of histidine to FIGLU. In support of this hypothesis, Arakawa *et al.* (1973) demonstrated decreased levels of histidase in DPH-treated rats.

Epileptic patients showing no evidence of folate deficiency in peripheral blood may still display signs of altered folate metabolism. Reynolds and Milner (1966) studied folic acid metabolism in treated but non-anaemic epileptic outpatients to determine the incidence of morphological changes in peripheral blood and bone marrow. They found that 38% of the patients showed megaloblastic haemopoiesis in bone marrow, but no morphological changes in peripheral blood were present. When seven of these individuals were treated with folic acid, normoblastic conditions were restored. The diet of all participants except one was judged to be good or adequate, including the folate intake. Duration of therapy did not seem to correlate with serum folate concentrations.

The folate concentrations of the cerebrospinal fluid (CSF) have also been found to be depressed in anticonvulsant-treated patients, and a good correlation exists between CSF and serum folate (Reynolds et al., 1971). The higher the drug levels, the lower the CSF folate.

The mechanism of action by which anticonvulsants precipitate folic acid deficiency has not yet been elucidated. It has, however, been tentatively suggested that anticonvulsant-mediated folate deficiency may be due to malabsorption and altered metabolism of folic acid.

Folate in foodstuffs is a mixture of monoglutamates and polyglutamates of varying chain length. Absorption is active, mainly in the upper part of the small intestine, and for absorption to be maximal a peptidase in the mucosa of the small intestine must split off the polyglutamate chain (Rosenberg and Goodwin, 1971). Earlier workers have suggested that anticonvulsants, DPH in particular, affect folate absorption by inhibiting intestinal peptidase (Hoffbrand, 1971). However, more recently, other investigators (Perry and Chanarin, 1972) have failed to confirm these findings. These workers have shown that monoglutamate forms are absorbed better than polyglutamate forms, but neither DPH nor bicarbonate inhibits the activity of intestinal peptidase.

According to Benn and co-workers (1971), the administration of folic acid in the monoglutamate form with either sodium bicarbonate or DPH results in lower serum levels of folic acid than when the vitamin is administered alone. These workers believe that the DPH, like the bicarbonate, increases the intraluminal pH of the gut, and that this effect is responsible for the impaired absorption of folic acid. However, this hypothesis has not been supported by Gerson (1971) who, by using a triple lumen perfusion system in nine healthy volunteers, investigated the effect of DPH on the absorption of tritiated pteroylglutamic acid. Although there was a significant decrease in its absorption when 20 µg/ml DPH was added to the perfusion solution, the intraluminal pH did not change.

There appears to be consistent agreement as to the fact that DPH impedes gastrointestinal absorption of folic acid, but the mechanism remains controversial. In recent years, much speculation has surrounded the possibility that induction of hepatic microsomal enzymes may be responsible for anti-convulsant-mediated altered metabolism of folic acid.

Folic acid may be required as a cofactor in hydroxylation of anticonvulsants. Thus, Furlanut and co-workers (1978) have shown that 14 days' folate therapy produces the greatest fall in serum concentrations of DPH in healthy individuals, and that such a fall is associated with increased urinary excretions of polar DPH metabolites. Folate therapy has also been shown to decrease plasma half-lives of phenytoin and antipyrine.

Furthermore, the antiepileptic agents are known to be potent inducers of NADPH-dependent hepatic drug-metabolising enzymes (Richens, 1972; Latham *et al.*, 1973). It is therefore possible that the folate-antagonistic effect of the anti-convulsants may be mediated through using up NADPH cofactor which is also required for the reduction of folate to its active forms, di- and tetrahydrofolate. Indeed, the degree of folate deficiency has been reported to be significantly related to increased hepatic microsomal enzyme activity as determined from an increased urinary excretion of *D*-gluconic acid in a group of 75 children with epilepsy (Maxwell *et al.*, 1972). The possible long-term effects of anticonvulsant administration on folate metabolism and its consequences are shown in Figure 3.2.

Correction of folate deficiency as a result of anticonvulsant therapy is an area of much debate. Folic acid treatment has been reported to provoke seizures in a few cases (Chanarin *et al.*, 1960; Wells, 1968), and fit frequency or severity has been found to increase in controlled trials (Reynolds, 1967; Bayliss *et al.*, 1971). In several other studies, no increase in fit frequency and/or severity was observed (Ralson *et al.*, 1970; Houben *et al.*, 1971; Mattson *et al.*, 1973), though duration of supple-

Figure 3.2: Possible long-term effects of anticonvulsants on folate metabolism

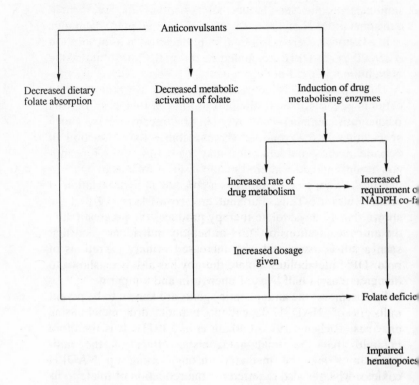

mentation may have been too short for any significant difference to be noticeable. This argument was presented by Reynolds (1973) on the grounds that a blood–brain barrier to folate transport prevents a significant rise in CSF folate level with folate treatment, even after 3 months (Spaans, 1970; Moir *et al.*, 1971). It has also been suggested that epilepsy may be exacerbated by folic acid therapy due to increased DPH metabolism and the resultant lowered drug levels in serum (Bayliss *et al.*, 1971).

Norris and Pratt (1971) carried out a controlled drug trial to measure fit frequency in 38 epileptic patients on anticonvulsant therapy. Of these patients 63% had borderline or frank folate deficiency. For the first 3 months they received placebos, and folic acid was administered for the subsequent 3 months. Since no change in number of fits was noted, they concluded that folate supplementation plays little or no role in seizure frequency. As the effects of folate deficiency may be quite serious, these findings support the treatment of this vitamin deficiency in epilepsy.

In a controlled double-blind trial of longer duration, Gibbard et al., (1981) instructed 57 patients, most of whom were taking only DPH, to record the number of seizures they suffered. Approximately half received a placebo, and the remaining patients were treated with folic acid for a minimum of 1 year. All patients had been on anticonvulsant therapy for a minimum of 6 months, although they were not selected for folate deficiency. Plasma drug levels were monitored, and an attempt was made to keep them consistent (10–15 µg/ml) throughout the trial. The results indicated that seizure frequency was actually less in those treated with folate, though the placebo group showed a decrease in seizure frequency. Consequently, the authors concluded that folate supplementation is probably safe for most anticonvulsant-treated patients.

In spite of the epileptogenic properties of folate compounds, Maugiere et al. (1975) claim that oral folate is compatible with full pharmacological control of epilepsy. Although it is theoretically possible that folates can cause seizures in very large doses (0.25 mg/kg: Chanarin et al., 1960; Hommes and Obbens, 1972), the dose that relieves folate deficiency in man is much less than this amount.

The importance of folate in pregnancy has been well established. Deficiencies of this vitamin increase the risk of congenital abnormalities, and seizures during pregnancy may result in infant and fetal morbidity (Janz, 1975). The risks of folate deficiency and seizures during pregnancy must be carefully weighed to determine if folate therapy should be instigated. Much criticism has been levelled against the indiscriminate use of folate therapy in pregnant epileptic women (Ralson et al., 1970; Mattson et al., 1973). If teratogenesis of anticonvulsants is dose-dependent (Harbison and Becker, 1972), lowering serum DPH levels to concentrations still affording full protection against seizures could decrease the risk of congenital malformation. Adjustment of dosage downwards may be justified during the critical twentieth to fortieth days of pregnancy.

In view of the possibility of aggravating seizures in some epileptic patients, folate should always be administered with caution. Its use may be justified if folate deficiency is present, especially if severe or prolonged, since there is evidence that a lack of folate may be detrimental to the nervous system (Reynolds, 1975) and to metabolic functioning. Indications for dosage and duration of treatment still need to be clarified.

3.6.4 Vitamin D

Osteomalacia due to vitamin D deficiency is recognised as a consequence of long-term treatment with anticonvulsants (Dent *et al.*, 1970). Vitamin D is first converted to 25-hydroxycholecalciferol (25-OH-D) in the liver, and then to 1,25-dihydroxycholecalciferol (1,25-$(OH)_2$-D) in the kidneys (see Figure 4.4). The latter is thought to stimulate the synthesis of a specific calcium-binding protein and calcium-dependent ATPase activity in the intestinal mucosal cells (Wasserman, 1970). It is through these actions that the metabolite 1,25$(OH)_2$-D is most actively involved in promoting intestinal calcium absorption and bone resorption (Lawson *et al.*, 1971). It has been suggested that prolonged anticonvulsant therapy increases the conversion of vitamin D in the liver to the more polar inactive metabolite instead of the active intermediate 25-OH-D (Stamp, 1974). In the treatment of anticonvulsive osteomalacia, it has been recommended that 50 000 IU/m^2 body area per week in divided doses plus 500 mg of supplemental calcium per day should be administered until the biochemical and radiographical parameters are brought to normal levels (Hahn and Alvioli, 1975).

3.6.5 Other vitamins

In addition to the vitamins discussed above, there are other vitamins which are also known to be affected by a wide variety of drugs through a number of mechanisms (Table 3.3). These mechanisms may include a direct effect causing morphological change in the mucosa of the small intestine, inhibition of mucosal enzymes, binding of bile acids, and interference with utilisation of vitamins.

Absorption of the fat-soluble vitamins has been reported to be impaired by the long-term administration of a large variety of drugs which include cholestyramine, mineral oil and broad-spectrum antibiotics. The lipid lowering agent, cholestyramine,

Table 3.3: Drugs affecting vitamin status

Drug	Vitamin	Possible mechanism
Cholestyramine	All fat-soluble vitamins; vitamin B^{12}	Binding of bile salts; inhibition of IF function

Antibiotics
  Neomycin
  Kanamycin
  Tetracycline
  Chloramphenicol
  Polymyxin
  Sulphonamides
	Vitamin K	Changes in gut flora

Neomycin Kanamycin	Vitamin B^{12}	Inhibition of IF function
p-Aminosalicylic acid Trifluoperazine Colchicine Phenformin	Vitamin B^{12}	Damage to the intestinal wall
Antacids	Vitamins A and B^1	Decreased absorption
Coumarins	Vitamin K	Antagonistic effect
Phenothiazine	Riboflavin	Antagonistic effect
5-Fluoromacil	Vitamin B^1	Competitive inhibition of thiamin phosphorylation

binds acids and prevents the solubilisation and absorption of fatty nutrients (Truswell, 1974). Instances of bleeding due to vitamin K deficiency (Visintine *et al.*, 1961) have resulted from a prolonged administration of the bile acid sequestrant.

Cholestyramine has also been shown to lower the absorption of radio-labelled B^{12} (Schilling test) in normal volunteers and in those with pernicious anaemia (Coronato and Glass, 1973). Using guinea pig intestinal mucosa, these workers have demonstrated that cholestyramine decreases the uptake of vitamin B12 by binding to the same sites on the intrinsic factor (IF) molecule which normally binds B12, thus preventing the formation of the IF–B12 complex.

Mineral oil, used as a stool softener, acts as a solvent for fat-soluble vitamins. Deficiencies of these vitamins due to prolonged use of mineral oil have been reported (Fingl, 1975).

Oral administration of broad-spectrum antibiotics, including neomycin, kanamycin, tetracycline, chloramphenicol and sulphonamides, may lead to vitamin K deficiency by inhibiting bacterial synthesis of the vitamin (Frick *et al.*, 1967). The antibiotic neomycin may also interfere with the action of bile salts and decrease pancreatic lipase activity and hence the absorption of fat-soluble vitamins, vitamin A in particular (Faloon, 1966).

Using the Schilling test, Jacobson *et al.* (1960) demonstrated that oral administration of neomycin to patients (12 g/day for one week) causes reversible vitamin B12 absorption. The mechanism of action is not known, but may be due to reduced IF function or to a direct influence on the mucosa of the small intestine (Faloon, 1970). Vitamin B12 deficiency may also result from treatment with a number of drugs either as a result of malabsorption of the vitamin or as a result of other known mechanisms. The drugs that impair the absorption of B12 without interfering with IF factors are *p*-aminosalicylic acid (PAS), colchicine and biguanides. Thus, using the Schilling test on 55 patients taking 12 to 16 g of PAS daily for pulmonary tuberculosis, Palva and his colleagues (1966) demonstrated that the test values were less than 10% in 28 patients. A study involving three obese healthy subjects receiving colchicine orally (1.9–3.9 mg/day for 4 to 8 days) showed a reduced B12 absorption, and that the absorption was reversible on withdrawal of the drug (Race *et al.*, 1970). Twenty-one out of 71 diabetic patients were found to show abnormally low B12 absorption following treatment with metformin for more than 2 years (Tomkin *et al.*, 1971).

A number of drugs are known to interfere with vitamin utilisation by blocking or altering transformation of the vitamin to its metabolically active form. Unlike drugs that interfere with vitamin absorption, agents acting by this mechanism usually produce relatively selective deficiencies. The effect of methotrexate (4-amino N^{10}-methyl pteroylglutamic acid, MTX) a widely used cytotoxic agent, on folate utilisation illustrates this mechanism well. In this example, the altered vitamin utilisation is of therapeutic significance. The enzymic activation of folate to dehydrofolate is inhibited by MTX because of the latter's close resemblance in structure to the vitamin (Osborne *et al.*, 1958), and restriction of cellular growth is attained by inhibiting nucleic acid synthesis in patients with malignant disease.

Coumarins, because of their close resemblance in structure to vitamin K, act competitively with the vitamin and reduce hepatic synthesis of the clotting factor, prothrombin. The altered vitamin utilisation seems to be of therapeutic significance. Sometimes, however, as a result of an increased dosage of coumarin, the prothrombin level may be decreased to the extent that haemorrhage becomes difficult to control. In this situation, sodium salt of vitamin K can be used as an antidote.

The biochemical evidence derived from both human and animal studies suggests that the widely used cytotoxic agent, 5-fluorouracil (5-FU) acts as a thiamin antagonist (Basu *et al.*, 1979; Aksoy *et al.*, 1980). However, the ways in which the antagonistic effect may be mediated are yet to be determined.

3.7 DRUG AND MINERAL INTERACTIONS

Although mineral elements constitute only a small portion (4%) of the total body weight, they serve as a foundation for body metabolism. These elements maintain the rigidity of the structural skeleton; control the water balance; regulate the acid–base balance of body fluids; act as a component of hormones, vitamins and enzymes; participate as catalysts in body reactions; and facilitate nerve impulses and muscular contractions. With the disruption of normal mineral metabolism, severe and often dramatic physiological and psychological reactions may occur.

Minerals can be classed as either macro- or microelements, depending on the amount of each that is needed in the diet. Calcium, phosphorus, potassium, chlorine, sodium, magnesium and sulphur are generally considered as macroelements, since they are needed in the diet at levels of 100 mg/day or more. On the other hand, iron, iodine, fluorine, zinc, copper, selenium, cobalt, chromium, manganese, molybdenum and silicon are often classed as microelements, because daily dietary requirements are no higher than a few milligrams.

A number of factors, such as deficiency in soil, water and plants, mineral imbalances, and inadequate dietary intake, have been associated with the occurrence of deficiency of an element in humans. Only recently, it has been realised that mineral depletion may result from drug-induced malabsorption or urinary losses. Indeed, a large variety of drugs are known to impair the metabolism of various inorganic elements.

3.7.1 Iron

The absorption of iron from the gut can be affected by antacids and tetracycline. The interaction between iron and antacids may be a problem for pregnant women whose iron requirements are higher than at any other time, while many of them depend on antacids to alleviate one of their common complications, acidosis. Another example of this kind is protein-energy malnutrition (kwashiorkor), which is often treated with milk as a source of protein, ferrous sulphate to correct iron deficiency, and tetracycline to correct secondary infections. The results of this combined therapy may be negligible, since a milk diet is known to impair the absorption of ferrous sulphate and the calcium in milk reduces the absorption of tetracycline by chelation (Poskitt, 1974). Furthermore, large doses of tetracycline have been reported to inhibit iron absorption in animals by interfering with protein synthesis in the gut wall (Greenberger, 1973). Among the antibiotics studied, cycloheximide has also been found to cause a significant reduction in net intestinal transport of iron in animals, possibly by inhibiting protein synthesis.

In a double-blind crossover study, it has been demonstrated that iron in turn may also affect the absorption of tetracycline (Neuvonen *et al.*, 1970). In this study the individuals were given a therapeutic dose of tetracycline either alone or in combination with 40 mg iron. Iron appeared to result in a reduction of blood levels of tetracycline and its derivatives. On the basis of these results, it has been suggested that it might be wise to withhold iron supplements when patients are receiving oral tetracycline preparations. Most tetracycline preparations are rapidly but incompletely absorbed from the gastrointestinal tract. The plasma concentration of antibiotics during therapy should not fall below a certain level in order to obtain the maximum bacteriostatic effect.

The hypolipidaemic agent, cholestyramine resin, has also been found to impair iron absorption. This was first claimed by Kniffen and his colleagues (1970), who observed that a patient treated with cholestyramine on a long-term basis developed iron-deficiency anaemia with erythropoietic protoporphyria. Subsequently, *in vitro* experimental studies have revealed that the resin impairs iron absorption by binding appreciable amounts of haem and non-haem iron, and that cholestyramine treatment affects the intestinal absorption of non-haem iron in normal, bled and iron-loaded animals (Greenberger, 1973). It therefore appears that the possibility of developing occult iron-deficiency anaemia as a result of long-term use of such agents by patients with a variety of disorders should not be ignored.

There are also drugs which may impair production of erythrocytes by either destroying erythrocyte precursors in the bone marrow or producing changes that deprive them of some nutrient that is required for erythropoiesis to proceed normally. Drugs that have been implicated in the production of aplastic anaemia include diphenylhydantoin (Yunis et al., 1967), meprobamate (Meyer et al., 1957), phenylbutazone (Fraumene, 1967), chloramphenicol (Wallerstein et al., 1969), and aspirin (Wijnja, 1966). Of these drugs, chloramphenicol is by far the leading offender (Best, 1967; Wallerstein et al., 1969).

In large doses, drugs such as phenylhydrazine may lead to haemolytic anaemia, which is characterised by the oxidative degradation of haemoglobin with its precipitation into particles of denatured protein (Heinz bodies). However, it should be pointed out that, whereas the normal red cell can be overwhelmed by large doses of drugs such as phenylhydrazine, the metabolic defences of other cells are well able to cope with a lesser insult. These defences consist primarily of the systems that maintain glutathione in its reduced form, and of glutathione peroxidase, which catalyses the detoxification of hydrogen peroxide by reduced glutathione (Figure 3.3). Furthermore, glucose-6-phosphate dehydrogenase is required to reduce NADP to NADPH in erythrocytes, and without a source of NADPH the red cells are unable to maintain glutathione in its reduced form. Subjects who have a hereditary lack of one of these enzymes (e.g. those of Mediterranean ancestry, and American negroes) may become susceptible to an oxidative challenge.

The red blood cell is generally capable of defending itself against the oxidative stress produced by administration of drugs such as primaquine and sulphonamides when the structure of haemoglobin is normal. However, haemoglobin molecules with structural abnormalities may readily lose their haem, and hence undergo oxidative denaturation (Jacob, 1970; Jacob and Winterhalter, 1970), especially when oxidative drugs such as sulphonamide compounds are administered (Frick et al., 1962).

69

Figure 3.3: Detoxification of drug-induced oxidation of haemoglobin

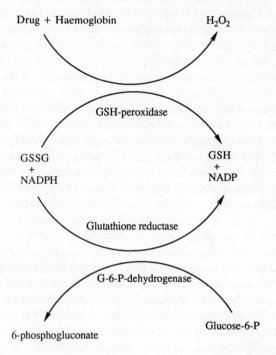

3.7.2 Electrolytes

Treatment with diuretics, purgatives and adrenal steroids has been found to lead to potassium deficiency by depleting the body of minerals. Most of these drugs are frequently used to treat aged people, whose dietary intake of potassium is often low to begin with (Davies *et al.*, 1973). Thus the elderly in particular are at risk of toxicity from a variety of drugs, and this applies particularly to digitalis which is known to be precipitated by a deficiency of potassium (Lown, 1956). Low total body potassium in kwashiorkor may also modify the metabolism of many drugs, digoxin being a notable example (Poskitt, 1974).

Digitalis along with a thiazide diuretic is frequently prescribed. Unless dietary potassium is increased to replace diuretic-mediated potassium losses, hypokalaemia occurs. Hypokalaemia enhances the effect of digitalis on myocardial irritability, and as a consequence dangerous ectopic arrhythmias may be potentiated (Mason, 1974). Therapy with potassium is usually beneficial for digitalis-induced arrhythmias (Chung, 1970; Selzer and Cohn, 1970).

A number of reports, however, have indicated that hypokalaemia may not always be associated with digitalis-induced arrhythmias (Dunbow and Burchell, 1965; Beller *et al.*, 1971), and therapy with potassium is beneficial for the condition. In order to shed some light on this seeming paradox, Brater and Morrelli (1977) have studied the electrophysiological effects of digitalis on the heart and intracellular potassium depletion during normokalaemia or hyperkalaemia. The results from this study indicate that digitalis-induced toxicity may be infrequently associated with hypokalaemia, but frequently associated with depleted intracellular potassium in the presence of either normokalaemia or hyperkalaemia. It therefore appears that the concentration ratio of potassium inside and outside the cardiac wall is more important than the plasma level of the electrolyte for prediction of the possible toxicity of digitalis. Careful monitoring of the electrocardiogram and serum electrolytes during administration of potassium may be of clinical significance since hyperkalaemia may also aggravate arrhythmias (Surawicz, 1968).

The irritant cathartics are derived from plants and exert their effects by stimulating intestinal motility. The most frequently used are derivatives of senna and cascara. Other substances in this group in common usage are castor oil, resins and phenolphthalein. Acute symptoms due to habitual use of these drugs are generally fatigue and thirst, which are characteristic of potassium deficiency. Treatment requires replacement of fluids, electrolytes (especially potassium), and gradual discontinuation of laxative habits.

Corticosteroids are known to possess salt- and water-retaining properties. Treatment with these drugs may therefore lead to weight gain with oedema and subsequently the development of hypertension (Dickerson, 1978). Similar side effects have also been found to occur in patients treated with phenylbutazone, oxyphenbutazone, carbenoxolone and oral contraceptive steroids.

3.7.3 Calcium

Many actions of calcium on the heart are similar to those of digitalis. Thus, an increase in the intracellular concentration of calcium ions can activate the contractile mechanism in a cell (Nayler, 1967). The administration of calcium therefore greatly enhances the action of digitalis (Nala *et al.*, 1970). High doses of calcium administered by intravenous infusion to dogs appear to enhance the sensitivity of the heart muscle to digitalis-induced arrhythmias (Nala *et al.*, 1970). The only clinical report of this drug–calcium interaction described two deaths that occurred following the intramuscular injection of digitalis and the intra-venous injection of calcium gluconate (Bower and Mengle, 1936). Although there is limited data about such interactions, care should be taken when administering intravenous calcium to patients receiving digitalis therapy.

Tetracycline is known to form insoluble chelates with metallic ions (Chin and Lach, 1975) including aluminium, magnesium and calcium ions. The antibiotic is absorbed through passive diffusion, and the formation of the metallic chelate may, in turn, depress the absorption of tetracycline. The ability to form these chelates, however, is dependent upon the pH of the gut; at low pH, there is little whereas, at high pH, high chelate formation occurs. Simultaneous administration of the metals or of products containing metals should be avoided in patients receiving tetracycline compounds.

3.7.4 Other inorganic elements

Treatment with a number of drugs, namely sulphonylureas, phenylbutazone, cobalt and lithium, has been reported to impair the absorption and utilisation of iodine, with a sub-sequent development of goitre (Dickerson, 1978). Other drugs which are known to interact with inorganic elements include oral diuretics, nephrotoxic drugs, penicillamine, and antacids containing aluminium hydroxide (Roe, 1984).

Magnesium and zinc are hyperexcreted by those receiving oral diuretics as well as by those on nephrotoxic agents. Penicillamine, which chelates metals, will induce depletion of zinc and copper. A phosphate depletion syndrome is known to occur in patients on heavy doses of antacids containing aluminium hydroxide (Ravid and Robson, 1976).

3.8 CONCLUSIONS

It therefore appears that a lack of dietary intake is not the only way vitamin and inorganic element deficiencies occur, and that a number of secondary mechanisms exist, one of which is drug-induced vitamin and mineral deficiencies. However, the level of significance of drug-mediated micronutrient deficiency in clinical and sub-clinical situations is often difficult to validate because of variables such as the nutrient status at the onset of therapy, the type of drug involved, the duration of treatment, and age. As a consequence, drug-induced micronutrient deficiency is a problem that is often overlooked. It is important that the probability of its occurrence as well as the clinical significance of many interactions are evaluated in the individual patient rather than by the use of generalised rules.

4

Alcohol and Nutrition

Alcohol differs from other pharmacologically active agents in that it is not generally used as a therapeutic agent by the medical profession. Alcoholic beverages are used by the consumer for their mood-altering effects and pleasing tastes. Although alcohol is generally socially acceptable, it may, even in moderate amounts, affect nutritional status. It is difficult, however, to define 'moderate' consumption of alcohol. Lieber (1984) considers 'moderate' a daily intake of up to 50 to 60 ml of absolute alcohol or 4–5 fl. oz. of 86° proof beverage. The most recent figures available on alcohol consumption show that it is increasing (Makela *et al.*, 1981). These figures parallel the rate of cirrhosis mortality (Table 4.1). It is estimated that in Canada the approximate number of hazardous drinkers increased from 260 000 in 1951 to 669 000 in 1970 (Alcoholism and Drug Addictions Research Foundation, 1972). In North America, alcoholism is now marked as either the third or fourth major health problem. The association between alcohol consumption

Table 4.1: Cirrhosis mortality rates per 100 000 population aged 25 years and over: International Study of Alcohol Control Experience (ISACE), 1950–1970

Country	1950	1955	1960	1965	1970	1975
California	26.2	29.0	33.3	39.4	40.3	36.3
Finland	3.7	6.1	5.8	6.0	7.4	10.1
Ireland	3.6	4.2	3.4	5.7	6.2	5.7
Netherlands	4.5	5.8	6.4	6.3	7.1	8.0
Ontario	7.7	9.7	12.9	13.9	18.0	22.2
Poland	—	—	6.1	10.9	15.2	18.0
Switzerland	—	21.1	19.2	25.0	25.2	20.4

Source: Makela *et al.*, (1981).

Figure 4.1: Relation between cirrhosis mortality rate per 100 000 population aged 25 years and over and alcohol consumption *per capita* in six Western countries (1950–1975)

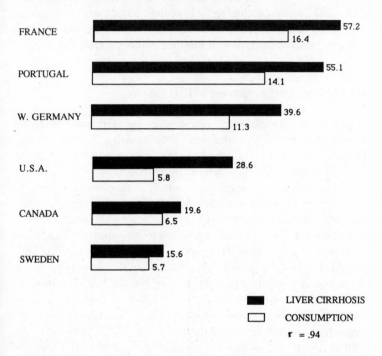

and hepatic cirrhosis has been known for centuries. The relation between the amount of alcohol consumed and the incidence of death from cirrhosis in various Western countries is shown in Figure 4.1

4.1 ALCOHOL METABOLISM

The principal active compound in alcohol beverages is ethyl alcohol (ethanol). Alcohol is rapidly absorbed across the biological membrane by simple diffusion. Complete absorption may require 2–6 h, depending upon the presence or absence of food in the stomach, the concentration of ethanol digested, the type of alcoholic beverages consumed, the rate of gastric emptying, and individual variations. As much as 95% of absorbed alcohol is metabolised in the liver, and the remaining

5% is excreted in the breath and urine (Gastineau *et al.*, 1979). In the liver, alcohol is metabolised at a rate of about 60–240 mg/kg/h, the average being 100 mg/kg/h. It is metabolised to CO_2 and H_2O, producing 7.1 kcal/g. There are two major routes of alcohol metabolism: the alcohol dehydrogenase system, and the microsomal ethanol oxidising system.

4.1.1 Alcohol dehyrogenase (ADH) system

ADH is a cytoplasmic, zinc-containing enzyme which is responsible for the production of acetaldehyde from alcohol (Figure 4.2). The acetaldehyde in turn is converted to acetate in a reaction catalysed by aldehyde dehydrogenase. The acetate is finally converted to acetyl CoA, which is then broken down to CO_2 and H_2O in the mitochondria. Both alcohol and aldehyde dehydrogenases require NAD^+ as a cofactor. The availability of unreduced NAD in the cell sap seems to be important in alcohol metabolism, since NADH competes with NAD for binding sites on ADH (Pawan, 1972). The rate of ethanol metabolism is therefore dependent upon the rate of $NADH^+$ reoxidation. In the cytoplasm of liver cells the reoxidation of $NADH^+$ may be coupled with the reduction of pyruvate to lactate, but the most important route for reoxidation of $NADH^+$ involves the mitochondrial flavoprotein–cytochrome electron transfer system coupled with oxidative phosphorylation.

4.1.2 Microsomal ethanol oxidising system (MEOS)

MEOS is an alternative pathway for the oxidation of ethanol to acetaldehyde (Figure 4.2). Its most striking feature is its utilisation of $NADPH^+$ to activate molecular oxygen; neither $NADP^+$ nor $NADH^+$ can replace $NADPH^+$. The systems which can generate NADPH are glucose-6-phosphate dehydrogenase, soluble isocitrate dehydrogenase, the malic enzyme producing pyruvate from malate, and the glutamate dehydrogenase producing α-ketoglutarate and NH_4^+. In addition to the MEOS, catalase, an enzyme located in the liver microsomes, can also catalyse the oxidation of ethanol to acetaldehyde, but only in the presence of H_2O_2. The overall metabolism of ethanol to acetate is summarised in Figure 4.2.

Figure 4.2: Metabolic pathways of alcohol

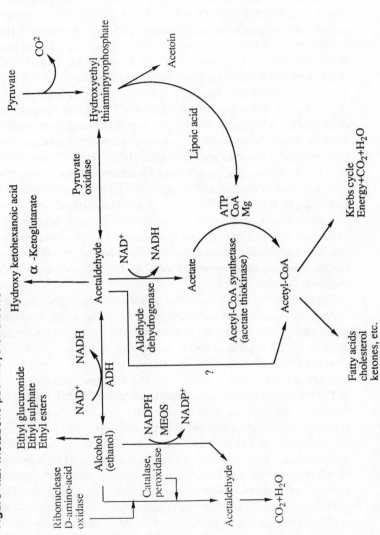

Source: Pawan (1972).

4.2 METABOLIC EFFECTS OF ALCOHOL IN THE LIVER

At one time it was thought that nutritional deficiency was solely responsible for the medical sequelae of alcohol abuse (Harttroft and Porta, 1968). However, it is now being realised that alcohol has direct toxic effects on a variety of tissues, including the gastrointestinal tract, the pancreas, the liver, bone marrow and the myocardium (Myerson, 1973); of these the hepatic toxicity of alcohol has been most convincingly established (Lieber and DeCarli, 1974, 1975). The alcohol-induced liver injury is due to the constant presence of alcohol itself and the metabolic derangements it causes. The chronic alcoholic with liver disease is usually also malnourished, which can further aggravate the hepatic injury (Lieber and DeCarli, 1977). The primary toxic effects of hepatic tissues can lead to malnutrition because the affected tissues cannot utilise nutrients or are unable to store nutrients in the physiological manner. The earliest and most common lesion of alcoholic liver disease is fatty infiltration. Most patients and experimental animals with fatty liver do not produce cirrhosis; however, some show a feature characteristic of alcoholic hepatitis which seems to be prognostic of later cirrhosis (Lieber, 1978).

Using electron microscopy the abnormal subcellular changes in the liver due to alcohol consumption can be seen. Some of these changes, for example hypertrophy of the smooth muscle, appear to be an adaptation to chronic drinking since this enhances the activities of the microsomal ethanol oxidising systems (Rubin and Lieber, 1971). Chronic ethanol consumption results in an acceleration of ethanol metabolism to acetaldehyde by the microsomal ethanol oxidising system, leading to an increase in the production and storage of acetaldehyde, a known hepatotoxic compound. In particular, acetaldehyde is known to cause mitochondrial impairment (Cederbaum et al., 1974). This is associated with a decreased capacity of hepatic mitochondria to oxidise acetaldehyde. The decrease of acetaldehyde metabolism in turn increases the acetaldehyde concentration in the liver, which may then perpetuate liver injury. Thus a vicious circle develops which may be harmful in alcoholics who indeed exhibit significantly higher levels of blood acetaldehyde compared with non-alcoholics given the same load of alcohol. It is possible then that hepatic injury due to chronic ethanol consumption can be attributed at least in part to increased levels of acetaldehyde.

Cederbaum *et al.* (1975) have studied the possible factors in the pathogenesis of ethanol-induced fatty liver. They have found that chronic ethanol consumption results in decreased fatty acid oxidation as shown by a reduction in oxygen uptake and carbon dioxide production. Also, chronic ethanol consumption leads to persistent impairment of mitochondrial oxidation of fatty acids to carbon dioxide. However, oxidation of fatty acids to AcCoA is not decreased by ethanol consumption. The metabolism of ingested alcohol results in an increased $NADH:NAD^+$ ratio in the liver (Lieber, 1973). The oxidation of free fatty acids is therefore slowed down as oxidation of the product, AcCoA, is limited by the rate of NAD^+-requiring steps in the citric acid cycle. Thus a greater proportion of hepatic free fatty acids are available for re-esterification to triglyceride. The increased $NADH:NAD^+$ ratio also raises the concentration of glycerol-3-phosphate, further enhancing the rate of esterification of free fatty acid to triglyceride. Also, excess NADH promotes fatty acid synthesis, possibly by the mitochondrial elongation pathway of transhydrogenation to NADP (Lieber, 1973).

Following alcohol ingestion an increased proportion of free fatty acids are incorporated into plasma triglyceride, and alcohol may induce, in liver microsomes, an increased capacity for lipoprotein synthesis (Nestel and Hirsch, 1965; Baraona *et al.*, 1973). Alcoholic lipaemia is characterised chiefly by hypertriglyceridaemia, but ethanol also induces cholesterol metabolism, there being decreased catabolism of cholesterol to bile salts (Lefevre *et al.*, 1972).

Fatty liver, hepatitis and cirrhosis are the most significant medical complications in the alcoholic in terms of morbidity and mortality. Approximately 85% of hepatic cirrhosis is due to alcohol, and 15–40% of heavy drinkers tend to develop cirrhosis, probably aggravated by individual susceptibility and malnutrition. Women appear to be more susceptible than men to the development of alcoholic cirrhosis (Wilkinson *et al.*, 1969).

4.3 ALCOHOL AND MALNUTRITION

Although the heat of combustion of ethanol is 7.1 kcal/g,

ethanol and other energy-rich dietary constituents do not behave as caloric equivalents. This is explained by the fact that isocaloric substitution of ethanol for dietary carbohydrates results in a decline in body weight (Pirola and Lieber, 1972). Ethanol metabolised by MEOS is a significant pathway especially when it is taken in large volumes in the absence of other food constituents. Since MEOS is not coupled to oxidative phosphorylation, ethanol metabolised by this system constitutes 'empty' calories. This may contribute to the weight loss observed in human beings upon isocaloric substitution of carbohydrate by ethanol (Pirola and Lieber, 1972).

Alcoholic beverages do not contain significant amounts of proteins, vitamins and minerals. The intake of these nutrients may readily become borderline or insufficient among chronic alcohol users. Furthermore, ethanol, either directly or through the action of its metabolic product, acetaldehyde, may affect the absorption and utilisation of vitamins and trace elements. In addition, ethanol may alter the metabolism of carbohydrates, lipids and proteins.

4.3.1 Vitamins

Thiamin (vitamin B1) deficiency is commonly found in alcoholics (Bonjour, 1980b). The deficiency is essentially due to its low dietary intake, impaired intestinal absorption, increased utilisation, and possibly depressed phosphorylation to the active form, thiamin pyrophosphate. Thiamin deficiency produces two disorders of the central nervous system, the neurasthenic syndrome and Wernicke's encephalopathy (Victor and Adams, 1961; Hell et al., 1976). The neurasthenic syndrome is characterised by lassitude, irritability and anorexia. Wernicke's encephalopathy, on the other hand, is associated with peripheral neuropathy and ataxia. All these signs may disappear with thiamin supplementations. However, there may be evidence of irreversible cerebral damage, depending upon the extent of thiamin deficiency.

The relationship between vitamin B6 status and ethanol metabolism has been extensively studied. A significant percentage of alcoholics appear to have an impaired status of this vitamin, as determined by its circulating levels as well as by the tryptophan-loading test (Bonjour, 1980a). The biochemical evidence of vitamin B6 deficiency seems to be caused by alcohol-mediated altered B6 metabolism. Thus alcohol or its oxidised product acetaldehyde may displace pyridoxal phosphate (the active coenzyme form of B6) from its cytosolic binding protein (Veitch et al., 1975). Such displacement may enhance its hydrolysis by pyridoxal phosphatase, resulting in a decrease in the activation of the vitamin. This would explain the finding that administration of ethanol to rats increases the urinary excretion of non-phosphorylated vitamin B6 compounds (Oura et al., 1963). Despite the high incidence of low circulating levels of vitamin B6 in alcoholics, overt clinical manifestations of the vitamin deficiency are rare.

Folacin deficiency is very common among alcoholics (Leevy et al., 1965; Eichner et al., 1971; Davis and Smith, 1974). Several factors may be responsible for the deficiency. Alcohol may decrease dietary intake and decrease absorption of the vitamin. It may also increase requirement and decrease hepatic affinity for folacin (Bonjour, 1979). Studies with volunteers have revealed that megaloblastic anaemia occurs only in subjects with an inadequate folacin store given a folacin-poor diet (Eichner et al., 1971), and this condition is accelerated by the intake of alcohol (Eichner and Hillman, 1971; Cowman, 1973). Megaloblastic anaemia is not precipitated, however, when subjects with adequate folacin stores are given a low-folacin diet (Eichner and Hillman, 1973) with and without alcohol (Eichner and Hillman, 1971). According to many surveys, folacin deficiency associated with megaloblastic anaemia is seen more frequently in malnourished than in well-nourished alcoholics (Hourihave and Weir, 1970; Williams and Girdwood, 1970).

Symptoms of vitamin A deficiency, such as night blindness and hypogonadism (McClain et al., 1979; Leo et al., 1981) as well as low circulating levels of the vitamin (Smith, J.C. et al., 1975; McClain et al., 1979; Leo and Lieber, 1982) have been described in alcoholics with liver damage. Dietary vitamin A is oxidised to active retinol by alcohol dehydrogenase, the enzyme which metabolises ethanol to acetaldehyde. This enzyme has a 50-fold greater affinity for ethanol than retinol. The possible mechanism for night blindness has been suggested to be the competitive inhibition of retinol formation by ethanol (Mezey and Holt, 1971).

The liver stores of vitamin A have also been reported to be

markedly depressed in subjects with all stages of alcoholic liver disease (Leo and Lieber, 1982). It is noteworthy that, despite the reduced hepatic stores of vitamin A, the serum concentrations of the vitamin appear to be normal in these subjects (Figure 4.3). In a recent report, vitamin A supplementation along with other vitamins has been recommended during detoxification therapy for alcohol withdrawal syndrome (Majunder *et al.*, 1983).

Figure 4.3: Hepatic and circulatory vitamin A in patients with chronic persistent hepatitis, alcoholic fatty liver, and alcoholic hepatitis

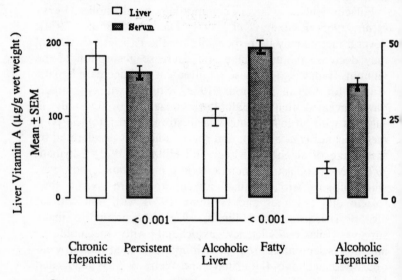

Source: Leo and Lieber (1982).

A history of alcoholism has been reported to be associated with the risk of hip fractures (Hutchinson *et al.*, 1979). There also appears to be a trend towards a greater risk of hip fractures in women as their alcohol consumption increases (Paganini-Hill *et al.*, 1981). Alcoholic men have been reported to have lower bone mass and to lose bone more rapidly than non-alcoholics (Nilsson and Westlin, 1973; Dalen and Lamke, 1976; Johnell *et al.*, 1982). Furthermore, regular use of alcohol has been suggested to be a contributory factor to an increased risk of fractures by predisposing such individuals to falls (Waller,

1978). The association between alcoholism and bone fracture could result from poor nutrition, reduced physical activity, or hepatic or chronic disease. In recent years, however, abnormal vitamin D metabolism has been suggested to be another possible factor.

Vitamin D from the diet or from the irradiation of 7-dehydrocholesterol in the skin accumulates in the liver where it undergoes the first metabolic alteration to 25-OH-D by a reaction requiring NADPH and molecular oxygen. The enzyme involved in this reaction, 25-hydroxylase, is believed to belong to the microsomal mixed-function oxidases, which metabolise

Figure 4.4: Possible interactions between alcohol and vitamin D metabolism

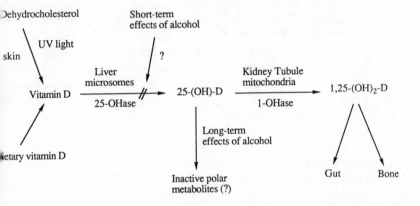

drugs. The resultant 25-OH-D is bound to α-globulin in plasma and transported to the kidney where it undergoes a second hydroxylation, forming $1,25\text{-(OH)}_2\text{-D}$ (Figure 4.4). It is this form of vitamin D which is believed to stimulate gut calcium transport, gut phosphate transport and bone calcium mobilisation. Reduction in serum concentrations of $1,25\text{-(OH)}_2\text{-D}$, due to either primary deficiency or metabolic derangement of vitamin D, could contribute to the loss of bone by decreasing absorption of calcium which, in turn, would increase secretion of parathyroid hormone (PTH). This increase in PTH accelerates the rate of bone loss (Parfitt et al., 1982).

The liver is important for both storage and metabolism of vitamin D, and interference with its functions by alcohol can therefore cause secondary nutritional deficiencies. It is well known that simultaneous administration of alcohol may potentiate the action of barbiturates and other drugs. The binding of alcohol to cytochrome P-450 has been suggested to inhibit competitively the metabolism of barbiturates (Ioannides and Parke, 1973). Furthermore, alcohol increases the NADPH-dependent MEOS activity (Lieber and DeCarli, 1970; Rubin *et al.*, 1970), resulting in further competition with the synthesis of NADPH-dependent microsomal mixed function oxidase system. For this reason, the ingestion of certain drugs (e.g. barbiturates, meprobamate and tranquillising agents) along with alcohol could be detrimental by virtue of the slow inactivation of the therapeutic agents. Taking this concept into consideration, it is not unreasonable to suggest that alcohol may also interfere with the hepatic microsomal 25-hydroxylase which is involved in the hepatic biotransformation of vitamin D to 25-OH-D. This hypothesis could be substantiated by the report by Avioli *et al.* (1967) that alcoholic cirrhosis appears to be associated with decreased clearance of vitamin D in the plasma and decreased urinary excretion of the vitamin conjugates.

Alcohol may also induce microsomal enzyme activities, provided the tissue concentration of alcohol is typical of chronic alcoholism for a sufficient period of time (Iber, 1977). The heightened activity of microsomal enzymes results in a faster metabolism of drugs when the individual is not burdened with alcohol. The alcohol-induced increase in hepatic drug metabolism generally lasts for 4 to 9 weeks after cessation of drinking. It is possible that a high tissue level of alcohol induces the production of enzymes in the liver which promote conversion of vitamin D to its more polar and inactive metabolites (conjugated vitamin D) rather than to its active metabolite, $1,25-(OH)_2-D$. Patients with chronic alcoholic hepatitis have been observed to have decreased serum 25-OH-D levels (Long *et al.*, 1976), and these results could be attributed to rapid urinary loss of the polar vitamin D metabolites (Krawitt *et al.*, 1977).

The above hypothesis in relation to alcohol-mediated altered vitamin D metabolism tends to concentrate on the hepatic enzyme, 25-hydroxylase. However, according to more recent reports, the principal effect of alcohol on vitamin metabolism tends to occur in the kidney where 25-OH-D is converted to either $1,25$-$(OH)_2$-D (active form) or $24,25$-$(OH)_2$-D (inactive form). Thus, alcohol administration to chickens has shown a decrease in the active metabolite and an increase in the inactive one (Kent *et al.*, 1979). Further studies have indicated that the formation of $1,25$-$(OH)_2$-D is under pituitary control (MacIntyre, 1979), the functioning of which is known to be affected by alcohol.

Whereas the mechanism by which alcohol affects the enzymes involved in vitamin D metabolism is not yet clearly understood, there appears to be no dispute concerning the alcohol-mediated altered metabolism of the vitamin. However, the extent to which this altered metabolism specifically contributes to clinical skeletal disorders in alcoholic populations remains to be elucidated.

Vitamin deficiencies occurring in alcohol abusers appear to be complex in aetiology. Causes of deficiency include dietary insufficiency, malabsorption, hyperexcretion, and a decreased rate of activation. Of all the vitamins, folacin and thiamin deficiencies are the most common among alcoholics. Deficiencies of these vitamins are often associated with clinical manifestations.

4.3.2 Inorganic elements

Iron absorption may be increased with acute alcohol intake as a result of the ethanol-induced gastric acid secretion (Darby, 1979). Furthermore, alcoholics may receive excessive iron from beverages, especially certain wines. It seems, therefore, that alcoholics are susceptible to increasing their hepatic iron stores, which could further aggravate alcohol-induced liver injury. Minerals such as zinc, calcium and magnesium may be decreased in the plasma of alcoholics as a result of decreased food intake, ethanol-induced renal losses, malabsorption and diarrhoea (Roe, 1981). Disorders of water and electrolyte balances, the most common being sodium and water retention, may be present in alcoholics with chronic liver disease. This is primarily due to the inhibitory effect of alcohol on the antidiuretic hormone. Clinically, salt and water retention may be manifested as oedema, ascites and pleural effusions.

4.3.3 Carbohydrates, lipids and proteins

Alcohol does not seem to have any effect on the absorption and digestion of carbohydrates, proteins and lipids. However, the metabolism of these nutrients can be affected by chronic alcohol intake. There is good evidence of increased hepatic fatty acid synthesis, decreased fatty acid oxidation, decreased citric acid cycle activity, decreased gluconeogenesis, and glycogen depletion (Tremolieres et al., 1972).

When alcohol is metabolised, NADH is produced in the cytoplasm and there is a noticeable increase in the ratio of NADH to NAD, which appears to affect the equilibrium of many reactions:

$$\text{Lactate} + \text{NAD} \rightleftharpoons \text{Pyruvate} + \text{NADH} \qquad \text{(Eq. 4.1)}$$

$$\alpha\text{-Glycerophosphate} + \text{NAD} \rightleftharpoons \text{Di-OH-acetone phosphate} + \text{NADH} \qquad \text{(Eq. 4.2)}$$

$$\text{Malate} + \text{NAD} \rightleftharpoons \text{Oxaloacetate} + \text{NADH} \qquad \text{(Eq. 4.3)}$$

Impairment of lactate oxidation (Eq. 4.1) leads to a depletion of pyruvate, the essential substrate for pyruvate carboxylase, which is the enzyme catalysing the first step in gluconeogenesis (Figure 4.5) from compounds which directly yield pyruvate, such as lactate, alanine or serine. Furthermore, when pyruvate is converted in the mitochondria to oxaloacetic acid (OAA), it then leaves the mitochondria as malate or aspartate, and must be converted back to OAA in the cytoplasm to serve as a substrate for conversion to phosphoenolpyruvate (PEP) and back to glucose. A high NADH/NAD ratio in the cytoplasm slows the conversion of malate to OAA (Eq. 4.3). Thus the efficiency of PEP carboxylase may be greatly reduced and the rate of gluconeogenesis decreased due to unavailability of OAA.

During alcohol metabolism, high concentrations of NADH may appear in mitochondria as a result of transfer from cytoplasm or from oxidation of acetaldehyde. High NADH can cause an accumulation of glutamate from the NAD-linked glutamate dehydrogenase reaction. High glutamate levels foster aspartate formation, thereby further limiting availability of OAA for PEP formation.

Figure 4.5: Metabolic pathway for gluconeogenesis

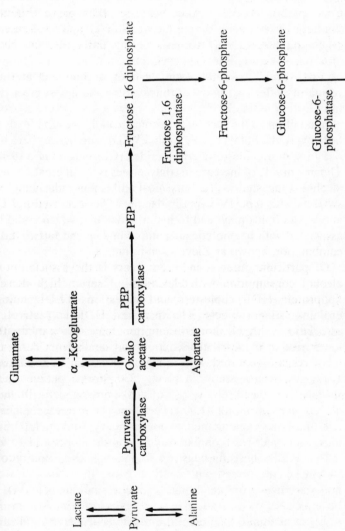

In the presence of alcohol, the hepatic concentration of α-glycerolphosphate (αGP) has been found to increase fivefold or more than when alcohol is not present (Krebs *et al.*, 1969). The increased level of αGP is also caused by the increased NADH/NAD ratio (Eq. 4.2) in the cytoplasm. Normally, fatty acyl CoA inhibits acetyl CoA carboxylase, thus preventing the synthesis of fatty acids during the times of lipid catabolism and regulating the rate of synthesis during times of anabolism. However, when αGP is abundantly present to act as an acceptor of fatty acyl CoA, the latter cannot accumulate and its regulatory effect is removed, thus causing the development of fatty liver (Lieber, 1967).

In addition to the fact that chronic alcohol ingestion leads to fatty liver, alcohol increases plasma VLDL and hence there is a rise in serum cholesterol triglycerides (Ginsberg *et al.*, 1974). The magnitude of increase in triglycerides is much greater when alcohol is ingested with a fatty meal and is particularly high in subjects with type IV hyperlipidaemia. The increased hepatic and serum lipids may lead to hyperlipidaemia, which could be associated with haemolytic anaemia, jaundice and fatty liver, a combination known as Zieve's syndrome.

Of particular interest in recent years is the association of alcohol consumption with elevations of serum high density lipoprotein (HDL) cholesterol and depressions of LDL cholesterol in fasting subjects. The increased HDL cholesterol in association with alcohol consumption may also explain the lower risk of myocardial infarction and death from coronary heart disease reported with moderate drinking (Yano *et al.*, 1977). A recent study involving 230 stroke patients also revealed a reduced relative risk of stroke among men with light alcohol consumption (10–90 g/week) and an increased relative risk (4.2 times) among men who were heavy drinkers (300 g or more per week) as compared with non-drinkers (Gill *et al.*, 1986). While these findings are interesting, the observations should be interpreted with caution. More than 90% of alcoholics are heavy smokers. Since cigarette smoking is known to be associated with coronary heart disease, drinking history *per se* is not adequate to predict the relationship between alcohol intake and heart disease.

The alcoholic often suffers from protein-energy malnutrition, essentially due to inadequate food intake. However, alcohol may block protein synthesis and alter the transport, uptake and clearance of amino acids. Experimentally, alcohol may be nitrogen sparing when given as supplementary calories, but decreases positive nitrogen balance when given as an isocaloric substitute for carbohydrate (Rodrigo *et al.*, 1971).

An increased accumulation of hepatic proteins (e.g. albumin, transferrin and glycoproteins), which are normally exported from the liver, has been found to be associated with chronic alcohol administration (Baraona *et al.*, 1977; Sorrell *et al.*, 1983). This accumulation of proteins is due not to increased synthesis but to a decrease in their secretion into the bloodstream. It has been suggested that acetaldehyde inhibits the polymerisation of tubulin, which is necessary for the production of hepatic protein secretions (Baraona *et al.*, 1977).

Chronic alcohol consumption has been implicated as a causative factor in gout associated with hyperuricaemia. Gout is generally caused by decreased urinary excretion and increased production of uric acid. Alcohol dehydrogenase promotes the production of NADH, which is used to form lactate from pyruvate. Thus, a high degree of alcohol metabolism might contribute to a lactic acidaemia. The resultant lactic acidaemia in turn suppresses the tubular secretion of uric acid, which is highly insoluble in its undissociated form (Na-urate crystals). Furthermore, recent studies have suggested that alcohol increases uric acid synthesis by stimulating turnover of adenine nucleotides (Faller and Fox, 1982).

4.4 ALCOHOL-RELATED CONDITIONS

The short-term effects of excess consumption of alcohol, being a sedative agent, include loss of inhibition, impaired co-ordination, reduced reflexes and mental function, and attitude changes. Long-term effects of regular heavy alcohol consumption could lead to gastritis, pancreatitis, cirrhosis of the liver, suppression of sex hormone production, and brain and nerve damage. In recent years, both prospective and retrospective epidemiological studies have suggested that chronic alcohol consumption may even be a cancer hazard (Tuyns, 1979; Pollack *et al.*, 1984). Thus, alcohol intake has been reported to be associated with an increased risk of cancer of the mouth, pharynx, larynx and oesophagus.

Alcohol intake during pregnancy may be highly detrimental

Figure 4.6: Effects of prenatal exposure to alcohols on birth weight of rat pups ($\overline{X} \pm$ SEM)

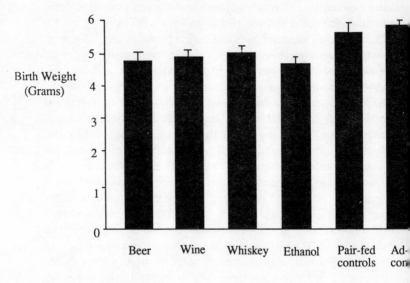

Source: Abel (1985).

to the unborn child. Since the fetal liver metabolises alcohol at half the rate of the mother, the alcohol retention in the fetal system will be twice as long. Women who drink heavily during pregnancy may have infants with fetal alcohol syndrome (FAS), which is characterised by physical, mental and behavioural abnormalities (Havlicek and Childaeva, 1976; Enloe, 1980; *Lancet* Editorial, 1983). There is also growing concern that moderate or even social levels of drinking during pregnancy may be hazardous to the fetus. FAS is believed to be the third most common cause of mental handicaps, after Down's syndrome and neural-tube defects (NTD). The most common fetal alcohol effect seems to be growth retardation during intrauterine life (Little, 1977; Abel, 1983). Regular alcohol exposure in experimental animals tends to reduce food and water intake, leading to a decreased maternal weight gain during pregnancy (Abel and Dintcheff, 1978). It is possible, therefore, that the decreased birth weight associated with prenatal alcohol exposure could be the reflection of alcohol-induced maternal undernutrition. The resultant undernutrition may, in turn, reduce the metabolism of alcohol, increasing its biological half-life (Wiener *et al.*, 1981). However, when control and experimental animals were fed a restricted amount of food and water in a pair-feeding fashion, the alcohol-exposed offspring still weighed significantly less than the control offspring (Figure 4.6). These results suggest that alcohol exposure *per se* could be contributory, at least in part, to intrauterine growth retardation. It also appears that the third trimester of pregnancy is the period in prenatal life that is most susceptible to alcohol exposure.

5

Possible Adverse Effects of the Pharmacological Use of Vitamins

The pharmacological use of vitamins can be divided into three areas: the prevention or treatment of diseases of deficiency, the treatment of vitamin-responsive inborn errors of metabolism, and the currently popular 'orthomolecular' megadoses. The 'orthomolecular' practice is used neither to correct a deficiency nor to treat a genetic abnormality in the absorption and utilisation of vitamins. Many arguments have been advanced in favour of the use of vitamins in large doses, both prophylactically and therapeutically (Evans and Lacey, 1986). The rationale of the use of megavitamins is to saturate the enzyme system and maximise the metabolic processes. It is estimated that about one-third of Canadians use vitamin supplements regularly at a cost of approximately \$47 per year (Griffith and Innes, 1983). Vitamins which are commonly taken in large doses include ascorbic acid (vitamin C), niacin, pyridoxine (vitamin B6), α-tocopherol (vitamin E), retinol (vitamin A), β-carotene (pro-vitamin A), and cholecalciferol (vitamin D) (Ovesen, 1986). The arguments about the beneficial effects of these vitamins in large doses have been countered by claims that most of the vitamins may be potentially toxic when taken in large quantities, through hepatic damage, neurological disorders and interaction with drugs which are simultaneously consumed (Alhadeff *et al.*, 1984). This chapter reviews the potential toxicity effects of vitamins which are commonly taken in large doses by seemingly healthy populations.

5.1 FAT-SOLUBLE VITAMINS

Among fat-soluble vitamins, hypervitaminosis with toxic mani-

92

festations is associated with vitamins A and D. In most instances, vitamins E and K have not led to adverse effects when consumed in pharmacological quantities over prolonged periods, although vitamin E may be antagonistic to vitamin K functions (Corrigan and Marcus, 1974; Korsan-Bengsten *et al.*, 1974). A list of commonly used brands of fat-soluble vitamins is given in Table 5.1.

Table 5.1: Brand names for fat-soluble vitamins containing mega quantities

Vitamins	Brand names
A	Acon, Afaxin, Alphalin, Aquasol A, Dispatabs, Sust-A
D	Calciferol, Deltalin, Drisdol, Ergocalciferol
E	Aquasol E, Chew-E, Eprolin, Pheryl E
K	Kappadione, Synkayvite, Mephyton, Aqua Mephyton, Konakion

5.1.1 Vitamin A and carotenoids

Vitamin A occurs physiologically as the alcohol (retinol), the aldehyde (retinaldehyde), the acid (retinoic acid) and the esters (retinyl esters). The major sources of vitamin A in the diet are certain plant carotenoid pigments such as β-carotene and the long-chain retinyl esters found in animal tissues. Carotenoids are converted to vitamin A primarily in the intestinal mucosa. Vitamin A is believed to be essential to the maintenance, general growth and differentiation of most epithelial tissues (DeLuca *et al.*, 1972). In vitamin A deficiency, cellular differentiation is altered, leading to keratinisation of epithelial cells. During this process, mucous membranes change from a single layer of mucin-secreting or ciliated epithelium to multi-layered keratinised epithelial cells resembling those of the skin. Therapeutically, vitamin A has been used for many years in the treatment of night blindness and xerophthalmia. In these conditions, 5000 IU/kg/day for 5 days has been recommended as the initial dose (Moore, 1960; McLaren, 1964). More recently, many reports have suggested massive oral doses

(200 000 IU) of vitamin A in an oil solution every six months to be effective as an emergency measure to prevent blindness due to primary vitamin A deficiency (Olson, 1972; Bauernfeind *et al.*, 1974).

Vitamin A has also been used to treat a variety of skin disorders, such as acne vulgaris, warts, Darier's disease, and skin cancer (Bollag, 1974; Orfanos and Schuppli, 1978; Levine and Meyskens, 1980; Pochi, 1982). Furthermore, both biochemical and epidemiological studies have provided evidence to support a link between vitamin A deficiency and cancer of epithelial cell origin (Basu, 1977b). Many studies to date suggest that the vitamin may have prophylactic as well as therapeutic activity against human cancer (Kummet and Meyskens, 1983).

The therapeutic use of vitamin A, however, has its limitations, as large doses of the vitamin may be potentially toxic (Olson, 1983). Adverse effects of vitamin A consumed from vitamin-A-rich foods are rare. The only outbreaks of acute toxicity that have been observed are among arctic explorers who ingest large quantities of vitamin-A-rich polar bear, seal or shark liver. Overconsumption of vitamin A preparations (15 000–50 000 retinol equivalent daily) over a long period can lead to chronic toxicity. The daily ingestion of vitamin supplements (some containing high vitamin A) to assure a full daily intake of essential nutrients is a growing practice in developed countries. On the other hand, in the developing countries, intermittent oral or intramuscular administration of massive doses of vitamin A is frequently practised in order to eradicate blindness due to vitamin A deficiency in children. Such practices may lead to hypervitaminosis A, which is characterised by anorexia, loss of weight, nausea, vomiting, drying and scaling of lips, and various types of skin rashes, muscular stiffness, enlargement of the liver, spleen and lymph glands, polydipsia and polyuria. The clinical manifestations resulting from altered cell function in deficiency and toxicity of vitamin A are shown in Figure 5.1.

The limitations of natural vitamin A being potentially toxic in large doses have led to the synthesis of a large number of new retinoids (Lasnitzki, 1976), based on the retinoic acid molecule with modifications of the ring, side chain and polar terminal group (Figure 5.2). It has been shown that the analogue 13-*cis*-retinoic acid binds freely to serum albumin and in this form can

Figure 5.1: Biological response to vitamin A deficiency, normalcy and toxicity in man and animals

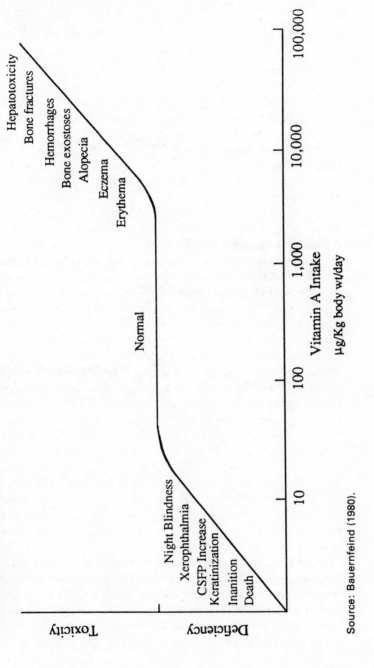

Source: Bauernfeind (1980).

Figure 5.2: Structures of new synthetic retinoids

All-trans-retinoic acid

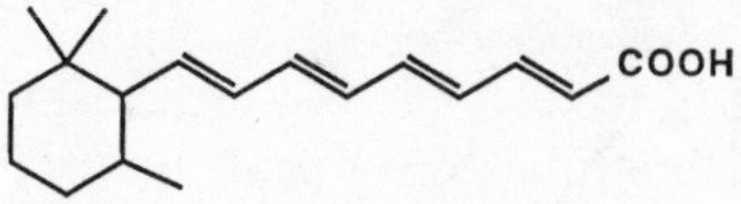

Cyclopentenyl analogue

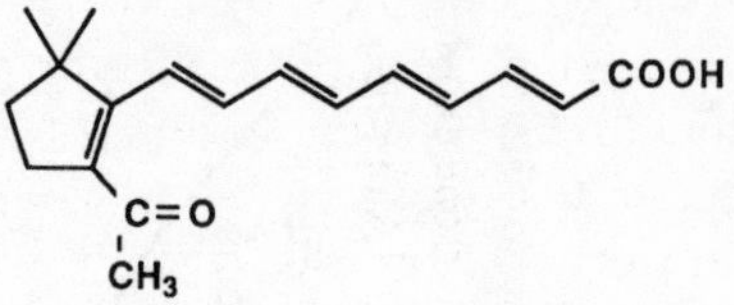

Aromatic analogues
R= -COOH
R= -COOC$_2$H$_5$
R= -CONHC$_2$H$_5$

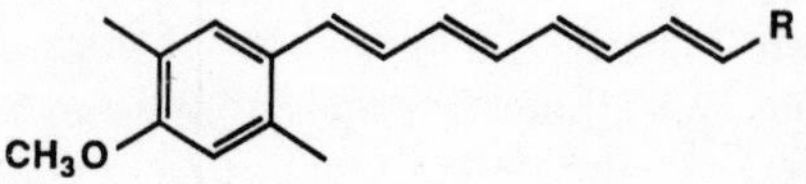

A$_2$-retinoic acid

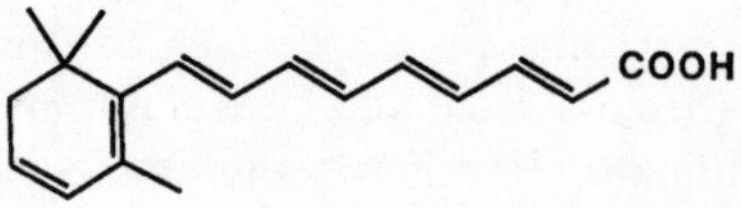

13-cis-retinoic acid

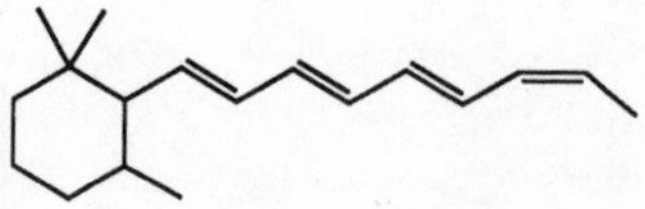

be readily distributed to target tissues by the bloodstream (Sporn *et al.*, 1976). No deleterious side effects were seen from prolonged feeding of the retinoid at doses as high as 9 mg/week (Port *et al.*, 1975). In contrast, an equivalent dose or even half the dose of retinyl acetate or all-*trans*-retinoic acid would have been severely toxic.

Vitamin A toxicity is essentially the result of misuse or over-consumption of the vitamin, either out of ignorance or careless-ness, or out of the misconception that excessively high intakes will provide some unusual health benefit. The incidence of death attributed solely to hypervitaminosis A is rare. In most instances hypervitaminosis A causes tissue changes, pain and discomfort, all of which are reversible upon reduction in vitamin A intake.

Although vitamin A toxicity remains a very minor clinical or nutritional problem, the possible teratogenic actions of the vitamin remain an overriding concern. Experimental evidence suggests that excess vitamin A given to pregnant females is teratogenic (Shenefelt, 1972). The administration of an excess amount of vitamin A (35 000 IU) to pregnant rats from the second to the sixteenth day post coitum produces a diminished litter size and gross cranial anomalies among the surviving young. The teratogenic action of vitamin A excess has also been shown in other mammalian species including guinea pigs, rabbits, pigs, hamsters and monkeys. Excess vitamin A in these animals during their pregnancies appears to produce a variety of congenital defects, such as cleft palate, eye defects, hydro-cephalus, spina bifida and cranial anomalies (Kalter and Warkany, 1961; Giroud, 1970).

Circumstantial evidence has indicated that maternal vitamin A excess mediated congenital abnormalities may also occur in humans (Bernhardt and Dorsey, 1974). In an attempt to assess the correlation between congenital malformations in infants with vitamin A intake of the mother, Gal and her colleagues (1972) measured the maternal serum vitamin A after delivery. The level of the vitamin was found to be significantly higher in mothers of infants with central nervous system malformations than in those with normal babies. Measurement of vitamin A in the livers of fetuses that were aborted, or of premature infants that died shortly after birth, were also higher in those with malformations. The relatively high vitamin A concentration found both in maternal blood and fetal liver suggests a possible teratogenic effect of excess vitamin A in humans.

The mechanism by which excess vitamin A leads to abnormal development of embryos and fetuses is not yet fully understood. It is thought that some of the effects of hypervitaminosis A may be due to labilisation of lysosomal membranes (Dingle and Lucy, 1965). More recent work indicates that the toxic manifestations of excess vitamin A occur when large amounts of vitamin A are presented to the cell membrane in association with lipoproteins, rather than specifically bound to retinol-binding protein (Smith and Goodman, 1976). Until more fundamental studies have been carried out to elucidate the role of vitamin A in human teratology, it is of paramount importance that a cautious attitude should be taken towards the administration of the vitamin to pregnant women. According to Bauernfeind (1980), the vitamin A intake for pregnant women should not exceed 10 000 IU (3000 μg retinol equivalent).

Unlike vitamin A (stored essentially in the liver), β-carotene is stored in adipose tissues. Hence the consumption of excess amounts of pro-vitamin A for long periods of time does not seem to cause toxicity, except for the benign expression of carotenaemia. In hypercarotenaemia the skin becomes yellow, and such yellowing is particularly evident on the palms and soles and around the nasolabial folds. Carotenaemia may occur due to excess intake of green leafy vegetables or carrots, citrus fruits and tomatoes over a long period of time. It may also occur in association with hyperthyroidism, nephrotic syndrome, diabetes mellitus, recurrent neutropaenia, hypothalamic amenorrhoea, and an inborn error of metabolism (Vakil *et al.*, 1985) where there is impairment of conversion of carotene into vitamin A.

5.1.2 Vitamin D

Vitamin D from the diet or from the irradiation of 7-dehydro-cholesterol in the skin accumulates in the liver where it undergoes the first metabolic alteration to 25-OH-D by 25-hydroxylase requiring NADPH and molecular oxygen (Bhattacharyya and DeLuca, 1974). The resultant 25-OH-D is bound to an α_2-globulin in plasma and transported to the kidney where it undergoes a second obligatory hydroxylation to form $1,25\text{-}(OH)_2\text{-}D$ (Fraser and Kodicek, 1970). It is this metabolite which is believed to stimulate transport of intestinal calcium and phosphate, bone calcium mobilisation and other functions attributed to the function of vitamin D (DeLuca, 1974).

We have known for years the beneficial effects of vitamin D therapy in nutritional rickets and osteomalacia. The recent discovery of $1,25\text{-}(OH)_2\text{-}D$ as an activate metabolite of vitamin D is proving invaluable to the treatment of human diseases.

These include: hypoparathyroidism, azotaemic osteodystrophy, autosomal recessive vitamin-D-dependent rickets, hepatic disorders and osteoporosis (Pierides, 1981).

✗ An excess intake of vitamin D can result in mobilisation of bound calcium in the skeleton and increases the serum calcium level. This calcium is taken up by soft tissues, especially the kidney, resulting in nephrocalcinosis and metastatic calcification of other soft tissues such as blood vessels, myocardium, lungs and skin. The symptoms of hypervitaminosis D include loss of appetite, gastrointestinal disturbances, head and joint pains, and muscular weakness. These symptoms are reversible, but, if uncontrolled, death may occur as a result of kidney failure. Toxic effects of the vitamin are seen in the adult when doses exceed 1000–3000 IU/kg/day after several months. In infants, hypercalcaemia may occur with total daily doses of 3000–4000 IU (Committee on Nutrition, Academy of Pediatrics, 1963). The sensitivity of individuals to an excess of vitamin D is quite variable so that it is not possible to state the minimum safety dose limit. Some individuals show profound toxicity to amounts only slightly above the daily recommended dose (Corstens et al., 1986). The syndromes of idiopathic hypercalcaemia of infancy and congenital supravalvular aortic stenosis may be examples of this toxicity (Seelig, 1969). It is possible that a little excess vitamin D intake along with large amounts of calcium is sufficient to cause hypercalcaemia.

$1,25\text{-}(OH)_2\text{-}D$ has been considered to be a steroid hormone (Kolata, 1975), the production of which is strongly feedback-regulated at the physiological level (see Chapter 4). The homeostasis of blood calcium and phosphate levels determines the rate of production of $1,25\text{-}(OH)_2\text{-}D$. It should be pointed out, however, that there is no feedback control over the hepatic production of 25-OH-D. There is evidence to suggest that this product may also be active in transporting calcium from gut lumen (Causins et al., 1970) especially when its concentration is increased. It is therefore possible that the toxicity caused by excess vitamin D administration is mediated through 25-OH-D rather than $1,25\text{-}(OH)_2\text{-}D$ (Hughes et al., 1976).

5.2 WATER-SOLUBLE VITAMINS

Unlike fat-soluble vitamins, water-soluble vitamins are readily excreted in the urine when there is excess in the body. Hence, in the past, megadoses of water-soluble vitamins were considered to pose little problem. However, in recent years concern has been expressed with regard to the potential hazards of prolonged and regular ingestion of large amounts of these vitamins. This section delineates the evidence supporting the fact that some of the water-soluble vitamins may not be as innocuous in large doses as is generally believed.

5.2.1 Vitamin C

During the last 15 years, there have been a number of reports claiming that vitamin C in large doses (100–1000 times the physiological dose) may have certain prophylactic and therapeutic effects in many conditions which include infectious diseases, immune deficiency disorders, atherosclerosis, malignant diseases, mental conditions and allergies (Basu and Schorah, 1982). The usefulness of large doses of the vitamin in treating or preventing these conditions is, however, exceedingly controversial. None the less, claims for megavitamin C being beneficial in various pathological conditions have stimulated widespread interest in self-medication with megadoses of the vitamin. This, in turn, led to several experimental and human studies evaluating the safety of high intakes of vitamin C. Some of these reports give cause for concern with regard to the potential hazards of prolonged ingestion of large amounts of the vitamin.

Megadoses of vitamin C have been reported to result in retardation of growth in experimental animals. Thus, Nandi and his associates (1973) demonstrated that the daily administration of 50 mg of vitamin C to guinea pigs fed a wheat flour diet resulted in a significant decrease in growth compared with that obtained on 5–20 mg of the vitamin per day, and was accompanied by 50% mortality in 16 days and 100% mortality in 25 days. With 100 mg of vitamin C, the effect was more drastic and all animals died within 16 days. Supplementation of this wheat diet by lysine, an essential amino acid found in only low concentrations in wheat, counteracted the toxicity of large doses of the vitamin.

It is of further interest that the administration of vitamin C (3 g/day for three successive weeks) to five healthy volunteers has been shown to reduce the urinary excretion of lysine to less than 50% of pre-vitamin C values (Basu, 1979). Thus it may be that the findings with guinea pigs (Nandi *et al.*, 1973) are applicable to human beings, so that, in communities where the diet consists mainly of cereals, intake of large doses of vitamin C may be harmful. Other workers (Sorensen *et al.*, 1974; Yew, 1973) also found that guinea pigs treated with megadoses of vitamin C had a reduced growth rate compared with animals fed a normal diet. There are, however, many studies that reveal no such effects (Hornig *et al.*, 1973).

The daily administration of 5 g ascorbate to healthy volunteers resulted in increased lysis of erythrocytes (Mengel and Green, 1976). These workers also found that patients with glucose-6-phosphate dehydrogenase deficiency are more susceptible to the toxic effect of megadoses of ascorbate. This is supported by the report of the death of a negro in which megadoses of ascorbic acid were a contributing factor (Campbell *et al.*, 1975). The individual concerned was a 60-year-old man with acute renal failure following second degree burns for which he had been treated with 80 g of ascorbic acid per day, intravenously, for 2 days. Following treatment with ascorbate, the patient became oliguric, with extensive haemolysis and urinary excretion of haemoglobin (4 mg per 100 ml). The patient's erythrocytes were found to be deficient in glucose-6-phosphate dehydrogenase, and he died on the 22nd day following ascorbate treatment.

There have been a few reports in the literature suggesting that vitamin C in large doses tends to favour infertility or abortion (Samborskaya and Ferdman, 1966). On the other hand, many workers have failed to observe such effects either in experimental animals (Basu, 1985) or among their patients taking as much as 10 g of vitamin C daily for many years (Poser and Smith, 1972; Wilson and Loh, 1973). Recently, Paul and Duttagupta (1978) investigated the effects of vitamin C in large doses (50 mg/day) on the reproductive organs in male rats fed either *ad libitum* or unrestricted diets. This study revealed that the restriction of diet alone reduced citric acid levels and the weight of reproductive organs, and that vitamin C therapy improved these effects. On the other hand, vitamin C treatment of the *ad libitum* fed rats resulted in a reduction in the weight and citric acid content of the male reproductive glands. These findings seem to indicate that an excess of vitamin C reverses the adverse effects of diet restriction on the male reproductive glands, while it induces the adverse effects in the *ad libitum* fed animals.

Vitamin C is partly metabolised to oxalic acid which is excreted in the urine. On the basis of this it has been theorised that the greater the quantity of oxalate excreted, the greater the probability of calcium oxalate calculi formation, and therefore the intake of vitamin C in large doses may lead to the formation of renal stones. A number of workers have cautioned that regular ingestion of large amounts of vitamin C may have adverse effects attributable to increased oxalic acid excretion (Lamden, 1971; Briggs *et al.*, 1973). However, the effect of vitamin C on oxalate excretion is subject to considerable individual variation. This may account for the fact that many workers have failed to note any significant increase in oxalate excretion on administration of up to 5 g daily of vitamin C for a long period of time (Murphy and Zelman, 1965; Takiguchi *et al.*, 1966).

Megadoses of vitamin C have also been reported to affect uric acid metabolism in man. Thus, Stein *et al.*, (1976) found that 4 g daily doses of the vitamin caused a two-fold increase in uric acid excretion and a decrease in blood urate concentration. These workers suggested that the vitamin-C-induced uricosuria was due to a decrease in the binding of urate to plasma proteins. Uric acid is the end product of purine base catabolism, and purines require amino acids for their formation. Increased uric acid excretion may therefore be a route by which vitamin C leads to depletion of body nitrogen. Furthermore, the excessive renal tubular uric acid may also lead to the precipitation of urate stones, especially as uric acid will be less soluble in the acid urine which will be produced during high vitamin C intake.

Long-term administration of large doses of vitamin C appears to lead to adaptation which might be responsible for vitamin C deficiency following a sudden cessation of the extra dietary vitamin intake (Schrauzer and Rhead, 1973). Two cases of infantile scurvy following maternal supplementation (400 mg/day) during pregnancy have been reported (Cochrane, 1965). There appears to be an increased need for vitamin C in the offspring of guinea pigs which have been given high doses of the vitamin throughout pregnancy and lactation (Nandi *et al.*, 1977). A more recent study has shown that the conditioning effect is less pronounced in guinea pigs when exposed to high vitamin C following weaning age than *in utero* (Basu, 1985).

In 1974, Herbert and Jacob reported that increasing levels of vitamin C added to homogenised meals prior to incubation at 37°C for 30 min as a laboratory mimic of the gastric environment in man produced increasing destruction of vitamin B12. However, these workers measured B12 in food by a method which was developed for assays of the vitamin in serum, where mild extraction appeared to be adequate. Since vitamin B12 in food, unlike serum, is usually tightly bound to proteins, it is essential that methods involving extensive extraction procedures which will release the bound vitamin are used for assaying the B12 content of food. Indeed, using one of these methods, Newmark *et al.*, (1976) showed that there were no significant deleterious effects of added vitamin C on vitamin B12 stability in foods. More recently, it has been reported that, among the cobalamins, including methyl-, hydroxy-, aquo- and adenosyl-cobalamin, only aquocobalamin is readily reduced and subsequently destroyed by vitamin C (Hogenkamp, 1980). It appears that large doses of vitamin C will not destroy all of the cobalamins. Use of megavitamin C has been reported to interact with many drugs if simultaneously taken. Thus, vitamin C may impair the response of the anticoagulant warfarin (Rosenthal, 1971; Smith *et al.*, 1972), increase the risk of drug crystalluria with aspirin (Meyers, 1972), and decrease renal tubular reabsorption of both amphetamines and tricyclic anti-depressants (Young *et al.*, 1975).

Vitamin C in excess also appears to depress microsomal drug metabolising enzymes as in its deficiency state. Thus oral administrations of the vitamin (300 mg/day) in guinea pigs for four successive days have been found to decrease significantly the specific activities of hepatic cytochrome P-450. Furthermore, a dose–response relationship of vitamin C with respect to the haemoprotein concentrations in guinea pigs has revealed that, at a daily intake of 50 mg for four days, there is a significant increase in cytochrome P-450 activity (Sutton *et al.*, 1983). However, the activity returns to control levels when the dose is increased to 100 mg/day, and a dose of 200 mg/day results in cytochrome P-450 falling below control levels. The decrease in activity is further exacerbated by 300 mg/day. The diminishing effect of high doses of vitamin C on the hepatic microsomal drug metabolising enzymes has been further supported by others (Ginter *et al.*, 1984).

Vitamin C in large doses also appears to impair the detoxification of cyanide by monopolising cysteine (Basu, 1977c, 1983b) which is required to metabolise vitamin C (as ascorbate sulphate) and to detoxify cyanide (to thiocyanate). It seems, therefore, that individuals taking megadoses of vitamin C concurrently with cyanogenic compounds (e.g. laetrile) may be subject to self-poisoning.

5.2.2 Vitamin B6

The term 'vitamin B6' refers to three pyridine derivatives which differ only in the functional group in the 4-position. These are the alcohol, pyridoxine, the aldehyde, pyridoxal, and the amine, pyridoxamine. All three compounds are metabolically interconverted in humans. In the form of pyridoxal 5'-phosphate (PLP), vitamin B6 acts as the coenzyme of a series of enzymes that catalyse transamination, decarboxylation, desulphydration, and tryptophan metabolism. The PLP-dependent decarboxylases convert amino acids to the corresponding biogenic amines (Table 5.2), some of which are substances of high physiological activity regulating blood vessel diameter and neurohormonal actions.

Table 5.2: Vitamin B^6-dependent synthesis of biogenic amines through decarboxylation of amino acids

Amino acids	Amines
Histidine	Histamine
5-OH-tryptophan	5-OH-tryptamine (serotonin)
Aspartic acid	p-Alanine
Glutamic acid	Gamma-aminobutyric acid (GABA)
Tyrosine	Noradrenaline
	Adrenaline
	Dopamine
Cysteic acid	Taurine

Because B6 is involved in the synthesis of amines, pyridoxine has gained public acceptance as a remedy for premenstrual syndrome (PMS) (Abraham and Hargrove, 1980; Gunn, 1985), carpal-tunnel syndrome (Ellis et al., 1982), schizophrenia (Joshi et al., 1982), hyperactivity, and learning disabilities (Brenner, 1982). As much as 2–6 g of vitamin B6 daily has been suggested for the treatment of these conditions, as opposed to the approximate daily physiological need for the vitamin of 1.5 mg.

Recent evidence indicates that large doses of vitamin B6 may exert a direct toxicity on the peripheral nervous system. The minimal toxic dose of vitamin B6 is believed to be 2 g. The clinical manifestations of toxicity consist of a tingling sensation in the neck and legs, and an unsteady gait (Schaumburg *et al.*, 1983). Although most of these symptoms disappear after withdrawal of the vitamin, varying degrees of residual neuropathy tend to remain for a very long time. Interestingly enough, like excess vitamin B6, a deficiency of this vitamin causes peripheral neuropathy. The neurological disorder probably results from a deficiency of biogenic amines. The mechanism by which excess vitamin B6 intake results in neuropathy is, however, far from being understood. Since pyridines are known to be neurotoxic (Sahenk and Mendell, 1980), the pyridine molecule of vitamin B6 may be a contributory factor to peripheral neuropathy, especially when the vitamin is taken in excess.

5.2.3 Niacin

Nicotinic acid and nicotinamide possess the same vitamin activity with different pharmacological and toxicity profiles. Unlike nicotinamide, nicotinic acid in large doses causes peripheral vasodilatation through releasing histamine. High doses (3 to 6 g/day) of nicotinic acid have been suggested for the treatment of a variety of conditions which include schizophrenia (Ridges, 1973; Hoffer, 1977) and hyperlipidaemia (Langer *et al.*, 1972; Coronary Drug Project Research Group, 1975).

Nicotinic acid, due to its histamine-releasing effect, may be contraindicated in patients with a peptic ulcer (Wentzler, 1979) and asthma (Ivey, 1979). Large intakes of the vitamin for a long period of time have been reported to precipitate hepatotoxicity (Dipalma and Ritchie, 1977; Ivey, 1979), cholestatic jaundice (Sugerman and Clark, 1974; Einstein *et al.*, 1975), and hyperglycaemia (Coronary Drug Project Research Group, 1975; Ivey, 1979). Because of the vasodilating effect, a single large dosage of nicotinic acid may stimulate heat sensation, itching, nausea, vomiting and headache. The vasodilating effect, however, wears off after several days of repeated administration.

Attempts have been made by the pharmaceutical industry to produce nicotinic acid derivatives which retain the hypolipidaemic effect but not its undesirable side effects. One of these analogues is β-pyridyl carbinol (Ronicol), which appears to be at least five times as potent in lowering fasting plasma cholesterol levels as is nicotinic acid (Marks, 1974).

5.3 CONCLUSIONS

Evidence to date concerning adverse effects of intakes of vitamins exceeding recommended amounts by large factors appears to be contradictory. Further studies are certainly warranted in order to clarify this conflicting situation through properly controlled studies. The nature of these studies should fall into three major categories: (a) the interaction of pharmacological doses of vitamins or their metabolites with other nutrients and drugs; (b) the basic mechanisms underlying deleterious effects of vitamins; and (c) the delineations of upper limits of vitamin intakes in quantities and dosing periods relative to possible benefits derived from large doses. Until these studies are carried out, it is of paramount importance to consider the reported adverse effects of megadoses of vitamins, albeit based on isolated cases. This is particularly important considering the fact that synthetic vitamins are readily obtainable in large amounts without prescription and that more and more people are taking the vitamins in orthomolecular amounts, and often along with other prescribed drugs. It is important for physicians to be aware of the toxic effects produced by individual vitamins either directly or through interaction with endogenous and exogenous compounds (e.g. drugs).

6

Drug–Food Interactions

The magnitude of pharmacological response to an orally administered drug is directly related to the extent and rate of gastrointestinal absorption of the drug. The absorption of drugs is influenced by many factors such as gastric pH, gut motility, and structure of the absorptive surface (Levine, 1970). The mixing of drugs with foods or beverages may alter these factors affecting the absorption through changes in their ionisation, stability, solubility, or transit time. Thus, food intake may increase the bioavailability of many drugs while it reduces that of others (Welling, 1977; Melander, 1978). The reduced absorption may result in prolonged drug action, and conversely when absorption is increased the blood level may be increased, perhaps to toxic levels. The clinical significance of drug–food interactions is reviewed in this chapter.

6.1 FOOD REDUCES/DELAYS DRUG ABSORPTION

The intestinal absorption of many drugs is reduced by con-current food intake essentially as a result of either delayed gastric emptying or dilution of the drug in the gut contents. It seems that gastric emptying time may be delayed by hot meals (Davenport, 1977), high fat content (Bachrach, 1959), and high-viscosity solutions (Levy and Jusko, 1965). Generally speaking, fatty meals delay gastric emptying time to a greater extent than either protein-rich or carbohydrate-rich meals (Bachrach, 1959).

Administration of a variety of drugs (Table 6.1) concomitant with food intake results in a delayed gastric emptying time. A

107

Table 6.1: Drugs whose absorption may be reduced/delayed by food

Digoxin	(D)	Penicillin V	(R)
Sulphonamides	(D)	Tetracycline	(R)
Acetaminophen	(D)	Amoxicillin	(D.R)
Cephalexin	(D)	Aspirin	(D.R)
Ampicillin	(R)	Levodopa	(R)
Erythromycin	(R)	Phenobarbital	(R)
Oxytetracycline	(R)	Isoniazid	(R)

(D) = delay; (R) = reduction.

comparison of the absorption of cephalosporins after oral doses administered to fasted subjects and to the same subjects 30 min after breakfast showed that the serum-drug profile was delayed in the non-fasted state (Harvengt *et al.*, 1973). These findings were later confirmed by others (Tetzlaff *et al.*, 1978). Delayed absorption of common drugs such as digoxin, sulphonamides and acetaminophen has also been reported, especially when these drugs are taken at the same time as food (Macdonald *et al.*, 1967; Jaffe *et al.*, 1971). It should be pointed out, however, that delayed absorption of drugs brought about by food does not always imply that a lesser total amount of a drug is absorbed, but rather that the time for a drug to reach peak levels after a single dose is lengthened (Macdonald *et al.*, 1967), suggesting a clinical advantage. Most penicillins (Table 6.1) appear to produce lowered peak serum levels following postprandial administration (McCarthy and Finland, 1960; Welling, 1977; McCraken *et al.*, 1978). Studies of the absorption of ampicillin or amoxicillin, administered to groups of patients who were given various test meals, showed a marked decrease in serum levels of the drugs when compared with those found in the fasted patients (Welling *et al.*, 1977).

Decreased absorption of tetracyclines due to chelation with metal ions has been well documented (Braybrooks *et al.*, 1975; Chin and Lach, 1975). Thus the absorption of tetracyclines is reduced by the presence of Ca^{2+}, Mg^{2+}, Fe^{2+} or Al^{3+}, as a result of the formation of insoluble compounds in the gastrointestinal tract. Serum levels of tetracyclines have been reported to be decreased by more than 50% if they are taken with ferrous sulphate or milk (Neuvonen, 1976).

Most drugs must be either small enough to pass through membrane pores or be non-ionised and lipid soluble for their

Table 6.2: Drugs that are irritating to the gastric mucosa

Aspirin	Steroids
Iron salts	Nitrofurantoin
Potassium supplements	Phenylbutazone
Aminophylline	Reserpine
Metronidazole	Diphenylhydantoin

passive transport across the gastrointestinal mucosa. Drugs that are acidic in nature are non-ionised in an acid medium, and the basic drugs are non-ionised in an alkaline environment. Changes in pH may therefore affect the extent of absorption of a drug by altering the ratio of its ionised form to its non-ionised form. The absorption of pseudoephedrine, for example, is increased by aluminium hydroxide (Lucarotti *et al.*, 1972), because the drug exists in the non-ionised form in alkaline conditions. On the other hand, antacids delay the absorption of pentobarbital (Hurwitz and Sheehan, 1971), which is a weak acid. There are, however, a variety of drugs listed in Table 6.2 which are intrinsically irritating to the gastric mucosa (Pierpaoli, 1972). The irritating effect of these drugs can be minimised if they are taken either just before or immediately after meals. Mixing drugs with fruit juices and beverages to mask their disagreeable taste may affect absorption due to decreased gastric pH (Hartshorn, 1977). Whole milk is much less acidic than most juices and has a pH range of 6.4–6.8. The drugs listed in Table 6.3 should not be mixed in acidic beverages (Lambert, 1975).

Table 6.3: Drugs that are acid labile

Erythromycin	Phenmetrazine
Lincomycin	Pseudoephedrine
Ampicillin	Quinone
Cloxacillin	Tetracyclines
Penicillin G	Aspirin
Penicillamine	

6.2 FOOD INCREASES DRUG ABSORPTION

A variety of drugs (Table 6.4) are better absorbed if taken with food. The antifungal agent griseofulvin is absorbed faster in the presence of a fatty meal. Serum levels of this drug have been shown to be about double when the drug is taken with such meals as compared with ingestion in the fasting state (Crounse, 1961). The bioavailability of nitrofurantoin has been reported to be increased by 20–400% in the presence of food (Rosenberg and Bates, 1976). The food-associated increase in the absorption of the drug has been attributed to a delayed stomach emptying time, permitting greater dissolution of the drug (Bates *et al.*, 1974).

Table 6.4: Drugs whose absorption may be increased by food

Griseofulvin	Carbamazepine
Nitrofurantoin	Propoxyphene
Propranolol	Methoxsalen
Metoprolol	Phenytoin
Hydralazine	Diazepam
Spironolactone	

The influence of concomitant food intake on the bioavailability of hydralazine has been studied by many workers (Melander *et al.*, 1977a; Liedholm *et al.*, 1982). It appears that the bioavailability of the drug is enhanced if taken with food, and this is true in both rapid and slow acetylators. Most gastrointestinal absorption occurs directly from the lumen of the gastrointestinal tract, across the epithelial cell lining into the adjacent capillary network of the circulation. The rate of blood flow through the splanchnic capillary bed will therefore have a marked effect on the absorption of drugs. Food increases splanchnic blood flow and hence enhances drug absorption (Mao and Jacobson, 1970). Since the gastrointestinal absorption of hydralazine is complete (Talseth, 1977), the food-associated increase of hydralazine bioavailability could be a consequence of a food-induced increase in splanchnic blood flow. A similar phenomenon has been observed for a food-associated increase in the absorption of other drugs, for example propranolol (Melander *et al.*, 1977b; McLean *et al.*, 1978).

Using antipyrine as a drug, Shively and his colleagues (1981) identified the 'timing of meals in relationship to the time of medication' as an important factor for drug disposition. Thus, meals appear to decrease protein-binding of aminopyrine, leading to an increase in the apparent volume of distribution and saliva half-life of antipyrine, possibly because of its increased tissue uptake. However, food-associated increase in bioavailability of drugs such as diazepam has been found to occur without affecting protein binding (Kratilla and Kangas, 1977). Thus it appears that the presence of food may increase the availability of drugs by a variety of mechanisms which include greater dissolution, dispersion, protein binding, and splanchnic blood flow.

6.3 FOOD AFFECTING URINARY EXCRETION OF DRUGS

Changes in urinary pH may have a profound effect on the excretion rates of drugs. Broadly speaking, a drug in its non-dissociated form diffuses more readily from urine back into the circulation. In other words, the pharmacological action of acidic drugs (e.g. aspirin, phenobarbital) is prolonged in acidic urine, since a larger proportion of the drug is non-dissociated in acid urine than in alkaline urine. The opposite phenomenon holds for a basic drug such as amphetamine or quinidine (an anti-arrhythmic drug). Normal dietary intake alone does not produce the urinary pH changes necessary to alter a drug's excretion rate. However, concomitant consumption of excessive amounts of acid- or alkaline-ash foods (Table 6.5) with drugs that acidify the urine (e.g. ascorbic acid, mandelamine, ammonium chloride) or alkalinise it (e.g. bicarbonate, carbonic anhydrase inhibitors) may lead to untoward consequences in drug therapy.

The renal reabsorption of quinidine is markedly increased if

Table 6.5: Foods that produce an acid or an alkaline residue

Acid residue	Alkaline residue
Bread	Milk
Cheese	Cream
Eggs	Ice cream
Bacon	Almonds
Meat	Fruit (except prunes,
Lentils	plums and cranberries)
Corn	Green vegetables
Oatmeal	
Nuts (except almonds)	
Macaroni	

this drug is taken with alkaline-ash beverages (Zinn, 1970; Embil *et al.*, 1976), leading to a serious arrhythmia (Lambert, 1975). The increase in urinary pH is believed to be responsible for such effects. On the other hand, acidic urine induced by ascorbic acid, for example, may decrease the tubular reabsorption of tricyclic antidepressants and amphetamines.

The urinary excretion rate of sulphisoxazole appears to be decreased during fasting, but the rate of biotransformation is unchanged (Reidenberg, 1977). Fasting has been found to decrease urine flow rate as well as to decrease pH. Both of these urinary changes could account for an increased non-ionic diffusion of sulphisoxazole from the renal tubular fluid back into the circulation, thus decreasing the excretion rate.

6.4 ADVERSE REACTIONS CAUSED BY FOOD

There are foods that contain active substances that can cause a direct drug-like effect, or that can interact with drugs to increase or decrease the expected response. Such reactions vary greatly in intensity, some being unpleasant and others life-threatening.

One of the most frequently discussed interactions is that which occurs between tyramine-containing foods and monoamine oxidase (MAO) inhibitors (see also Chapter 3). The MAO is the enzyme responsible for intracellular degradation of catecholamines. The inhibitors of the enzyme are antidepressant drugs, which include isocarboxazid, nialamide, pargyline-HCl, phenelzine sulphate, and tranylcypromine sulphate. These drugs elevate the concentrations of dopamine, serotonin and norephinephrine in brain and other tissues.

MAO inhibitors are contraindicated in patients who consume foods high in tryptamine or tyramine (see Table 3.1). Tyramine is a substance that resembles norepinephrine pharmacologically, and produces an elevation of blood pressure. MAO inhibitors prevent the metabolism of tyramine, which normally occurs in the gastrointestinal tract and liver. This results in an elevated blood level of unmetabolised tyramine, which may lead to raised blood pressure.

Cheeses that have been allowed to mature have a high tyramine content. The syndromes characterised by hypertension of short duration, headaches, palpitations, nausea and vomiting have been reported in patients receiving MAO inhibitors along with cheese (Asatoor *et al.*, 1963) or other tyramine-containing foods (Hodge *et al.*, 1964; Blackwell *et al.*, 1965; Nuessle and Norman, 1965; Rice *et al.*, 1976).

Other drugs that inhibit MAO include isoniazid (an anti-tubercular drug) and procarbazine (used in Hodgkin's disease). Tyramine reactions have been reported in patients receiving these drugs with food containing high levels of tyramine (Spivack, 1974; Kent and Durack, 1978).

Natural liquorice contains glycyrrhizinic acid, which is functionally similar to desoxycorticosterone. Because of the glycyrrhizinic acid content of liquorice, its excessive intake may lead to hypokalaemia, and to salt and water retention. Liquorice intake in excess is therefore contraindicated in patients receiving cardiac glycosides, thiazide diuretics and low-salt diets (Pierpaoli, 1972; Carmichael, 1973). Cardiac arrest as a result of liquorice-induced hypokalaemia has been reported in a patient consuming 1.8 kg of liquorice sweets per week (Bannister *et al.*, 1977).

Caffeine is a stimulant of the CNS. Excessive consumption of caffeine-containing beverages such as coffee or tea may affect the clinical effectiveness of neuroleptic drugs (Rippere, 1981). Coffee, because of its xanthine content, may also induce side effects of the bronchial dilator theophylline, especially when it is taken in large amounts.

Foods such as Brussels sprouts, cabbage, cauliflower, turnip, and soya beans contain a goitrogenic agent, thiooxazolidine, which is a potent inhibitor of the thyroid hormones (Lehmann, 1978). These foods are contraindicated in patients who are receiving thyroid medications.

Monosodium glutamate, a widely used food additive, has been found to produce a syndrome characterised by facial pressure, headache, chest pain and burning sensations of the extremities in susceptible individuals, especially if taken on an empty stomach. This syndrome is often referred to as 'Chinese restaurant syndrome' because of the frequent use of this substance in preparing Chinese food.

Normally, the glutamic acid synthesised in the body is more concentrated in the CNS than any other amino acid, and gamma-amino butyric acid (GABA), the decarboxylated product of glutamic acid (Figure 6.1), is found almost exclusively in nervous tissues. Glutamic acid is an excitatory amino acid, and

Figure 6.1: Metabolism of glutamic acid

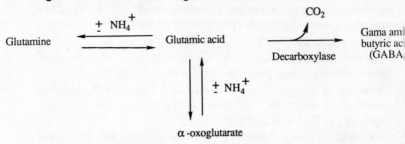

GABA has an inhibitory effect on the electrical behaviour of the nervous system. The properties of the two acids make them candidates for natural transmitters. It is possible that the adverse syndrome of administered glutamic acid in certain individuals is based on the disturbance of an equilibrium involving glutamic acid and its metabolic products glutamine, ammonium and α-oxoglutarate (Figure 6.1), on which the proper functioning of the nervous system may depend.

7

Nutrition and Experimental Carcinogenesis

Food may be related to the causation of cancer in various ways. The most obvious is the fact that food may contain potentially carcinogenic compounds which either occur naturally or are added by accident or intention during processing. A variety of mutagens are present in substantial quantities in fruits and vegetables, carcinogens are formed in cooking as a result of reactions involving proteins and lipids, and new chemicals in food are still being identified (Ames, 1983). Although these compounds are normally submitted to very stringent tests to make certain that none of them is potentially carcinogenic, there are some compounds (Table 7.1) which are still believed to be carcinogenic though this has been demonstrated only in animals. It might be argued that much of this experimental evidence has little direct relevance to human cancer. However, studies of the distribution of different types of cancer between and within populations have provided clues to possible links between food and oncological disease in man.

It has long been known that diet could modify the incidence of cancer. Much of the evidence comes primarily from geographic correlations, prospective studies of selected populations, dietary histories of cancer cases compared with suitable comparisons, and migrant studies. The most difficult question has been the identification of the specific dietary factors involved. A number of dietary components have been suggested, including a lack of dietary fibre, high fat consumption, high meat intake, vitamin deficiency, and inorganic element deficiency. Some of these nutrients, especially vitamins, selenium and dietary fibre, may be regarded as protective factors against cancer when present in adequate amounts in the

Table 7.1: A list of possible carcinogens of food origin

Carcinogens	Site of Tumour	References
Acetylaminofluorene	Liver, breast, small intestine	Miller *et al.*, 1958; Makiura *et al.*, 1974
Yellow rice toxins	Liver	Miller and Miller, 1976
Bracken fern	Small intestine, bladder	Pamuken *et al.*, 1971
Cycasin (methylazoxy-methanol β-glucoside)	Liver, kidney	Miller and Miller, 1976
Aflatoxins	Liver	Swenson *et al.*, 1975
Oestrogen	Breast, uterus	Herbst *et al.*, 1971
DDT (1,1,1-trichloro-2,2-bis-*p*-chlorophenyl ethane)	Liver	Tomatis *et al.*, 1972; Walker *et al.*, 1973
Safrol (1-allyl-3,4-methylenedioxybenzene)	Liver, oesophagus	Long *et al.*, 1963; Concon *et al.*, 1979
Cyclohexylamide	Bladder	Shubik, 1975
Nitrosamines	Liver, stomach	Issenberg, 1976
4-Methylaminoazobenzene	Liver	Makiura *et al.*, 1974
Hydrazines	Lung	Toth *et al.*, 1982
Gossypol	Skin	Haroz and Thomasson, 1980

diet, possibly by modifying the carcinogenic property of a variety of environmental chemicals. The protective effects of dietary fibre against cancer, especially of the colon and breast, have been discussed in Chapter 1. This section delineates the anticarcinogenic effects of micronutrients with particular references to vitamin A, β-carotene, vitamin E, vitamin C and selenium.

7.1 VITAMIN A (RETINOL)

The value of vitamin A in protecting against the premalignant condition squamous metaplasia has been shown in experimental animals exposed to carcinogenic polycyclic aromatic hydrocarbons (PAH). Thus, Saffiotti and his colleagues (1967) have demonstrated that intratracheal administration of benzo(*a*)-pyrene to Syrian golden hamsters induces squamous metaplasia in the tracheobronchial mucosa, and that oral administration of vitamin A palmitate following the benzopyrine treatment markedly inhibits metaplastic changes in squamous epithelia as well as the development of squamous tumours. Similar observations have been made using organ cultures of hamster tracheas showing that vitamin A inhibits squamous metaplasia induced by benzo(*a*)pyrine (Crocker and Sanders, 1970). More recently, Cone and Nettesheim (1973) have shown that vitamin A protects rats against tumours in response to a carcinogen, 3-methylcholanthrene, given by endotracheal instillation. Supplemental vitamin A has been further reported to prevent cancer of the forestomach and cervix in hamsters treated with PAH (Chu and Malmgren, 1965) and to inhibit growth of skin papillomas induced by 7,12-dimethylbenz(*a*)anthracene in mice (Davies, 1967). Newberne and Suphakan (1977) have shown that in rats exposed to dimethylhydrazine (DMH), all vitamin A deficient animals developed colon tumours compared with 60% of those supplemented with the vitamin. Other studies have shown that vitamin A deficiency increases the susceptibility of animals to induced tumours of the oral cavity (Rowe and Gorlin, 1959), lung (Nettesheim *et al.*, 1975) and bladder (Cohen *et al.*, 1976). It has been reported that vitamin A deficiency results in an elevated ornithine decarboxylase (ODC) activity in the colon mucosa of rats (Daliam *et al.*, 1986). Furthermore, administration of a bile acid appears to lead to a significantly larger ODC response in vitamin A deficient animals compared with non-deficient rats. These results suggest that a deficiency of vitamin A increases sensitivity to the promoter action of bile acids.

Vitamin A and its synthetic analogues have been reported to act in both the 'early' (initiation phase) and 'late' (promotion phase) stages of tumour induction, phases that correspond to the period of carcinogen metabolism and target cell interaction, and tumour development and growth, respectively (Hill and Grubbs, 1982; Sporn and Roberts, 1984). Hyperplastic and anaplastic epithelial lesions induced by chemical carcinogens can also be reversed by supplementation with vitamin A, even after the lesions have developed (Thompson *et al.*, 1979; McCormick *et al.*, 1980). Vitamin A is thus able to cause cellular repair of the neoplastic process induced by chemical carcinogens.

Although the mechanism of anticarcinogenic action of vitamin A remains to be elucidated, several possibilities exist. Vitamin A has been found to inhibit *in vitro* the microsomal mixed function oxidases that metabolise carcinogenic PAH (Hill and Shih, 1974). This may be one of the mechanisms involved in the protecting effect of the vitamin against PAH-induced squamous metaplasia. Another plausible explanation for the protective effect against PAH may be provided by the work of Genta *et al.* (1974), who showed that vitamin A deficiency enhances the binding of benzpyrene to hamster tracheal DNA. About four times as much carcinogen appears to bind tracheal DNA from hamsters fed a diet deficient in vitamin A as compared with normal animals.

Vitamin A is believed to stimulate post-translational synthesis of glycoproteins (Wolf *et al.*, 1979), which are involved in a variety of cellular functions including mediation of cellular adhesion and growth. Indeed, recent evidence has shown that vitamin A alters the agglutinability (Lotan, 1980) and the adhesive properties of transformed cells (Dion *et al.*, 1978).

Decarboxylation of ornithine is the rate-limiting step in the generation of polyamines, which are known to control the synthesis of nucleic acids and proteins (Russell and Durie, 1978). These substances are intimately involved in cell proliferation. Ornithine decarboxylase (ODC) activity correlates with cell proliferation and can be stimulated by tumour promoters such as bile acids. Taken together these findings suggest that vitamin A and its analogues may be involved in the modulation of carcinogenesis through a variety of mechanisms, including influences on tissue differentiation, cell kinetics, host immune functions, and a variety of biochemical pathways.

Smoking cigarettes is an important causative factor for lung cancer in humans. Both prospective (Bjelke, 1975; Hirayama, 1979; Shekelle *et al.*, 1981) and case control (Mettlin *et al.*, 1979; Gregor *et al.*, 1980) studies with lung cancer patients point to an inverse relationship between vitamin A intake or its serum levels and lung cancer risk. Furthermore, subnormal vitamin A status has been reported among smokers (Fehily *et al.*, 1980). Thus, there appears to be a link between smoking, lung cancer and vitamin A. The experimental and epidemiological evidence linking vitamin A and cancer is very important and has been extensively confirmed. However, we should prudently note (1) that the antitumour activity is tissue-specific (Graham, 1984); (2) that some tumours are actually stimulated, rather than inhibited, by vitamin A treatment (Smith, D.M. *et al.*, 1975); (3) that many studies have failed to demonstrate an inverse relationship between serum vitamin A and cancer cases (Wald *et al.*, 1974; Lambert *et al.*, 1981; Nomura *et al.*, 1985; Harris *et al.*, 1986); and that frequently dietary total vitamin A (i.e. retinol + β-carotene) rather than vitamin A alone is measured.

7.2 β-CAROTENE

In 1981, Peto *et al.* reviewed various data and hypothesised that β-carotene may be the major molecule providing antineoplastic activity. This sparked much active interest, and numerous reports have now appeared shining considerable new light on the question. The most consistent evidence of protective effects of high intakes of carotene-rich foods (green and yellow vegetables) comes from studies of lung cancer (Shekelle *et al.*, 1981; Ziegler *et al.*, 1984; Wu *et al.*, 1985). It may also be protective against cancer of other sites, such as stomach, oesophagus, ovary, cervix and breast (Hennekens *et al.*, 1986). Other evidence that β-carotene prevents cancer has come from animal studies. Thus the carotenoid has been found to have protective effects in rat, mouse and hamster against tumours at various sites (Table 7.2). All of these studies provided strong evidence

Table 7.2: Protective effects of dietary β-carotene against chemically induced tumours of various sites in animals

Dietary β-carotene (mg/kg)	Carcinogen	Tumour site	Species	References
33 000	DMBA[a]/UV	Skin	Mouse	Mathews-Roth, 1982
25–250	DMBA	Submandibular salivary gland	Rat	Alam and Alam, 1985
200	DMBA/croton oil	Skin papilloma	Mouse	Muto and Moriwaki, 1984
90	DMBA	Not specified	Rat	Rettura *et al.*, 1983
20	DMH	Colon	Mouse	Temple and Basu, 1987
Topical	DMBA	Buccal pouch	Hamster	Suda *et al.*, 1986

[a] DMBA = 7,12-dimethyl-benz(*a*)anthracene.

that β-carotene either delayed tumour appearance or reduced tumour yield. Although most studies have used very high levels of β-carotene, one study (Temple and Basu, 1987) has shown that the carotenoid in low dosage has a strong inhibitory action against 1,2-dimethylhydrazine- (DMH) induced colon tumours in mice. The dose of β-carotene used in the study was 20 mg/kg diet, which is equivalent to about 150–300 g carrots per 3000 kcal and is therefore in the nutritionally relevant range in humans. Such a low intake of β-carotene resulted in reductions of incidence and multiplicity of DMH-induced tumours by

Table 7.3: The anti-tumour activity of dietary β-carotene against DMH-induced colon tumour in mice[a] (Temple and Basu, 1987)

Dietary β-carotene (mg/kg)	No. of mice	Tumour incidence (%)		Tumour multiplicity (colon tumours/mouse)			
		Adenoma	Adenocarcinoma	Adenoma		Adenocarcinoma	
2	31	61.2	32.3	2.0	1.7	0.57	0.7
22	32	37.5	3.1*	1.2	0.4**	0.08	0.3*

[a] Mice were sacrificed 31 weeks after first does of DMH.
*$p = < 0.05$; **$p = < 0.01$.

about half. Adenomas, the predominant type of tumour, were about 40% less common in supplemented mice, and adeno-carcinomas were largely absent (Table 7.3). These observations are based on mice sacrificed 31 weeks after the first DMH injection. Observations of these animals for 13 additional weeks demonstrated that, consistent with the above data, β-carotene-supplemented mice had only half the mortality of unsupplemented mice (Temple and Basu, 1987).

An important question is whether β-carotene is most effective at the initiation or promotion stages of carcinogenesis. The best evidence comes from the hamster study which showed β-carotene to be clearly effective at both stages (Suda et al., 1986). Similarly, β-carotene was shown to inhibit DMBA-induced transformation of mouse mammary cells in vitro with activity apparently occurring at both stages (Som et al., 1984). Other studies have indicated that β-carotene can be effective when given only before the carcinogen (Seifter et al., 1984), only after it (Rettura et al., 1983) or only after tumours are already present (Muto and Moriwaki, 1984). The experiments using an oncogenic virus (Seifter et al., 1982) and a transplantable tumour (Rettura et al., 1982) point to a late-stage effect. Overall, β-carotene appears to block the initiation, promotion and subsequent development of tumours.

Although the precise mechanism of anti-cancer activity of carotenoids remains to be highlighted, several hypotheses have been proposed. It has been suggested that β-carotene may have an antioxidant action, particularly at the relatively low oxygen partial pressures found in most tissues under physiological conditions (Burton and Ingold, 1984). In other words, carotenoids are free-radical traps and efficient quenchers of singlet oxygen, which is particularly effective at causing lipid peroxidation (Foote, 1982). Carotenoids are believed to be the plant's main defence against singlet oxygen generated as a by-product from the interaction of light and chlorophyll (Krinsky and Deneke, 1982).

Another possibility is that β-carotene alters the metabolism of carcinogens so that they are preferentially detoxified. The dose–response effects of β-carotene on hepatic microsomal drug metabolising enzymes support this hypothesis. Thus the hepatic concentrations of cytochrome P-450, as well as the activity of hydroxylase using biphenyl as a drug substrate, have been found to be reduced in mice by feeding a diet containing β-carotene 20–500 mg/kg for 14 days (Basu et al., 1987). In parallel with this finding, it also appears that β-carotene treatment may lower the DMH-induced mortality rate.

In recent years, many investigators have suggested that β-carotene may be involved in stimulating the immune defence mechanism. This view is based on the fact that it increases the cytotoxicity of macrophages towards hamster tumour cells (Schwartz et al., 1986), enhances thymic function, particularly lymphocyte production (Seifter et al., 1982), and influences human interferon action (Rhodes et al., 1984).

Since β-carotene is a pro-vitamin A, it is also possible that the anticarcinogenic action of β-carotene reflects its vitamin A activity. However, the rate at which β-carotene is converted to vitamin A is determined by the presence of preformed vitamin A (Brubacher and Weiser, 1985). In view of the fact that, in most experimental studies, the diets were adequate in preformed vitamin A, it is doubtful that enough β-carotene was converted to vitamin A to appreciably raise the total intake. It is therefore plausible that the anticarcinogenic action of β-carotene is independent of its pro-vitamin A role. Furthermore, the carotenoid canthaxanthine, which is not a pro-vitamin A but is a good trapper of oxygen singlet, has been shown to protect hairless mice from tumours induced by UV-B irradiation (Mathews-Roth, 1985). It seems that there is much need for human and experimental studies to include more carotenoids not convertible to vitamin A, so that a clear distinction can be made between preformed vitamin A and carotenoids with regard to their anticancer properties.

121

7.3 VITAMIN E (α-TOCOPHEROL)

Vitamin E is another antioxidant in the diet that could be important in protecting body fat and lipid membranes against oxidation. The protective action of the antioxidant against lipid peroxidation is fundamental to the structural and functional integrity of the cell. Since the cellular membranes of all the organelles are lipoprotein, lipid peroxidation can break the membrane and thereby disrupt the membranous enzyme system. The generally accepted mechanism of peroxide formation is shown in Figure 7.1. A fatty acid molecule is first

Figure 7.1: Mechanism of vitamin E in trapping free radicals

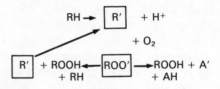

RH = FATTY ACID (UNSATURATED);
R′ = FREE RADICAL;
ROO′ = PEROXIDE;
AH = VITAMIN E.

oxidised, generating a free radical which, through reacting with oxygen, produces peroxide. The peroxide can further react with another fatty acid molecule to produce another free radical, and thus the reaction chain is propagated (Witting, 1974). There are a variety of foreign chemicals (Table 7.4) that can react with cellular lipid molecules and accelerate the generation of damaging radicals (Pryor, 1982). Vitamin E, due to its antioxidant property, is known to protect cellular lipid molecules from their chain auto-oxidation (Figure 7.1). It is the major radical trap in lipid membranes, and has been used clinically in a variety of oxidation-related diseases (Bieri *et al.*, 1983). Vitamin E has been shown to inhibit chemically induced cancers in a variety of species (Shamberger, 1970; Kurek and Corwin, 1982; Shklar, 1982). The vitamin is believed to reduce the activation of various carcinogens to epoxides, which are more effective than the parent compounds in producing malignant transformation (Walton and Packer, 1980). Protective effects of vitamin E against radiation-induced DNA damage and DMH-induced carcinogenesis have also been reported (Cook and McNamara, 1980; Beckman *et al.*, 1982). Vitamin E can also inhibit both *in vitro* and *in vivo* formation of nitrosamines by irreversibly reacting with nitrosating agents to form α-tocopherol quinone (Mergens *et al.*, 1979; Newmark and Mergens, 1981).

Table 7.4: Foreign substances that are responsible for generating free radicals

Foreign substances	Action
Soot, tar, tobacco, smoke, NO, NO_2	Contain free radicals
Ozone, singlet oxygen	Contribute to free radical formations in target molecules
CCl_4, nitrofuran drugs, bleomycin, paraquat	Form free radicals by interrupting cellular electron flow and single electron transfer reactions
Benzo(*a*)pyrene, dopa	Undergo auto-oxidation to form superoxide and hydrogen peroxide

Overall, vitamin E in large doses appears to have both antitumour and prophylactic activities against carcinogenesis in experimental animals (Prasad and Rama, 1984). Vitamin E blocks the action of certain tumour-promoting agents; it can also kill newly transformed cells directly or indirectly by stimulating the host's immune response and can reverse the malignant phenotype to a normal phenotype in certain tumours. It may also inhibit the production of prostaglandin E series, which are known to suppress the host's immune system. Although there appears to be a substantial amount of evidence indicating that vitamin E may have an influence on tumours induced in experimental animals, to date there is no epidemiological evidence that the vitamin has an influence on tumour induction in humans. The only remotely possible correlation has been through studies related to the serum levels of vitamin E. One of these studies failed to support the hypothesis relating serum levels of vitamin E to a reduced cancer risk (Willett *et al.*, 1984), and in another study the risk of lung cancer appeared to increase in a linear fashion with decreasing serum levels of the antioxidant (Menkes *et al.*, 1986).

7.4 SELENIUM

There is wide variation in dietary intakes of selenium in different parts of the world. The association between selenium deficiency and cancer incidence is currently one of the most complex, challenging and controversial areas of epidemiology. Several geographic studies have indicated that areas with higher soil and locally grown food contents of selenium experience lower cancer mortality rates (Shamberger and Willis, 1971; Clark, 1985; Combs and Clark, 1985). Furthermore, cancer patients often have a history of a low serum selenium level (Shamberger et al., 1973; Combs and Clark, 1985).

Of 37 animal studies recently reviewed, two-thirds reported that supplementary selenium caused a reduction in tumour incidence of at least 35% (Combs and Clark, 1985). The rat colon is one of the tumour models for which this has been well established (Birt et al., 1982; Jacobs, 1083; Reddy et al., 1985). With carcinogen-induced mammary tumours in rats and mice, it appears that selenium is protective at several stages of carcinogenesis and particularly against early promotion (Vernie, 1984). One of the plausible mechanisms of the anticarcinogenic action of selenium is thought to be mediated through inhibiting activation and accelerating detoxification of carcinogenic substances (Schrauzer, 1984). Selenium is also an antioxidant (Ames, 1983). It is required to activate and synthesise glutathione peroxidase, an enzyme essential for destroying lipid hydroperoxides and endogenous hydrogen peroxide (Figure 7.2) and thus helping to prevent free radical-induced lipid peroxidation (Flohe, 1982). There appears to be a selenium dose-dependent increase in the cellular oxidised glutathione (GSSG) to reduced glutathione (GSH) ratio both *in vitro* (Leboeuf and Hoekstra, 1985) and *in vivo* (Leboeuf and Hoekstra, 1983). This increase in GSSG has been found to be associated with an increase in the activity of protein kinase (Ernst et al., 1978). It seems possible, therefore, that the antitumour activity of selenium may also be mediated through inhibition of protein synthesis (Kosower et al., 1972) caused by the increased cellular level of protein kinase. Moreover, selenodiglutathione (GSSeSG), a reductive intermediate of selenite, has been shown to inhibit elongation of peptide synthesis by inactivating eukaryotic elongation factor 2 (Vernie et al., 1975).

Figure 7.2: Relationship between vitamin E and selenium in detoxifying peroxides

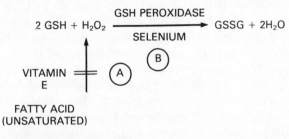

A = FIRST LINE OF DEFENCE
B = SECOND LINE OF DEFENCE

7.5 VITAMIN C (ASCORBIC ACID)

Patients with malignant disease have been reported to have minimal tissue stores of vitamin C (Krasner and Dymock, 1974). Both plasma and leucocyte levels are decreased in patients with leukaemia and pulmonary, skin, breast and prostatic cancers (Basu et al., 1974; Kakar and Wilson, 1974, 1976). Examinations of both primary and secondary tumours in man and experimental animals have shown that vitamin C is actively concentrated in malignant tumours at the expense of the surrounding tissue (Moriarty et al., 1977). There also appears to be some limited data suggesting that vitamin C and its oxidation products can inhibit the formation of certain carcinogens. Thus ascorbic acid, dehydroascorbic acid and 2-3-diketoglutonic acid have been shown to inhibit growth of the transplanted solid sarcoma-180 in mice, the highest inhibitory effect being shown by dehydroascorbic acid (Yamafuji et al., 1971; Omura et al., 1974). The inhibitory effect of vitamin C seems to be further enhanced by the addition of copper ions as copper sulphate, suggesting that the active agent might be an oxidation product of ascorbic acid. The synergistic effect of copper and vitamin C in inhibiting tumour growth is further evident in an *in vitro* study (Bram et al., 1980) involving malignant melanoma cells which are known to be rich in copper (Ikonopisov, 1972).

Several hypotheses have been advanced to explain the anti-tumour activity of vitamin C in the presence of copper ions.

125

Copper ions react with vitamin C and generate free radicals in solution (Peloux *et al.*, 1962) which, in turn, lead to reductions in the viscosity of DNA solutions (Yamafuji *et al.*, 1971). It has also been suggested that dehydroascorbic acid functions as an electron acceptor in the regulation of mitosis (Edgar, 1970).

Prevention by vitamin C of the development of tumours at specific sites has also been suggested. Thus, high doses of the vitamin in mice were reported to reduce colo-rectal tumour development from rectal polyps in susceptible inbred strains (Logue and Frommer, 1980) and also to reduce the number of potentially premalignant polyps in the human with familial polyposis (DeCosse *et al.*, 1975). Cancer of the bladder may be caused by an oxidation product of 3-hydroxyanthranilic acid, which is a derivative of tryptophan. It has been suggested that vitamin C prevents production of the carcinogen because of its strong reducing properties (Schlegel, 1975).

N-Nitroso compounds constitute an important group of environmental carcinogens. The widespread occurrence of these compounds in the environment is due to their formation from commonly occurring precursors, such as nitrite, amines and amides. Since nitrosation involves the interaction between a nitrogenous compound and a nitrosating intermediate, it is believed that any agent capable of rapidly reacting with the nitrosating intermediate and converting it into a non-nitrosating form may inhibit nitrosation. A role has been postulated for vitamin C in inhibiting the nitrosation formation (Kyrtopoulos, 1987). This topic has been discussed in Chapter 1.

The antitumour and anticarcinogenic activities of vitamin C have prompted many workers to test the therapeutic significance of the vitamin in human cancer. Thus Cameron and Campbell (1974) gave 10 g of vitamin C to terminal cancer patients and reported an extended survival period. Criticism of the non-random nature of the controls in the study was met by a further report (Cameron and Pauling, 1978), which again showed a palliative effect with matched controls. However, a subsequent random prospective double-blind study (Creagen *et al.*, 1979) of 150 patients with advanced cancer showed no therapeutic effect of the vitamin.

Although available evidence indicates that supplemental vitamin C could produce substantial benefits in both the prevention and the treatment of cancer, the subject still remains controversial. Nevertheless, reports of the use of vitamin C in the treatment of cancer have tentatively suggested that it may play a role in the body's immune process. Research on the effects of vitamin C on host defence, however, has proceeded in a piecemeal fashion and as a result the literature is bedevilled by controversy and lack of confirmation (Anderson, 1981).

7.6 CONCLUSIONS

Specific nutrients such as vitamin A, β-carotene, vitamin E, vitamin C, selenium, dietary fibre and fat (see Chapter 1), total calories, and alcohol (see Chapter 4) appear to have some relation with cancer. Some of these nutrients, especially vitamins, selenium and dietary fibre, may be regarded as protective factors against cancer when present in adequate amounts in the diet. On the other hand, excessive intake of dietary fat, total calories, and alcohol, may predispose an individual to developing the disease. It is conceivable, therefore, that one can exert a certain amount of control over one's own cancer risk through modifying diet and lifestyle.

The interrelations between nutrition and cancer are, however, subject to controversy, both within the scientific and medical community and among the general population. Controversies are inevitable when data are neither clear-cut nor complete. We have not yet been able to identify, at least with confidence, a particular food or nutrient that will make a striking difference in cancer risk reduction. All we have are the data that are suggestive of a causal relationship between dietary habits and cancer. These data are derived from correlation studies, international and national, and from case-comparison studies. Several experimental studies using animals and cancer-promoting substances (carcinogens) have made similar observations. On the basis of the available evidence, however inconclusive, it is important that we have dietary guidelines which may lower cancer risk. In 1982, the Committee on Diet, Nutrition and Cancer (USA) put forward such guidelines. These were subsequently (in 1984) translated by the American Institute for Cancer Research into a practical food guide. The essence of the guidelines is not only to lower cancer risk but also to reduce the risk of heart disease. The recommendations are as follows:

(1) Studies of the geographical distribution and experiments in

animals provide convincing evidence that increasing the intake of fat of both animal and vegetable origin increases the incidence of cancer at certain sites, especially the breast and bowel. Conversely, the risk is lower with a lower intake of fat. The recommendation, therefore, is to reduce total fat intake to a level of 30% of total calories.

(2) Fruits and vegetables (green and yellow) are good sources of vitamins and inorganic elements which appear to have protective effects against cancer. The protective effects may be mediated by promoting the health of the body cells, the immune defence mechanism, and the ability to detoxify foreign substances present in food which may have cancer-promoting potential. The consumption of fruits and vegetables should therefore be increased.

(3) The importance of dietary fibre in protecting against bowel diseases, including cancer, has received a great deal of attention. Although the role of dietary fibre as a protector against cancer is still a controversial issue, it is recommended that fibre intake should be increased by consuming fruits and vegetables, and by substituting unrefined whole grain products (oats, wheat, rice) for refined sugars and flours.

(4) Certain cooking methods, such as charcoal broiling, smoking and exposure to pyrolysing heat, may generate cancer-producing substances (i.e. polycyclic aromatic hydrocarbons). Ingestion of such substances with highly salted foods may be a contributory factor in causing cancer at some sites, especially the oesophagus and the stomach. It is therefore recommended that the consumption of salty, smoked or charcoal-broiled foods be minimised.

(5) Heavy consumption of alcoholic beverages appears to be associated with cancer of the mouth, pharynx, larynx and oesophagus. This relationship is obscured by the fact that most heavy drinkers are also smokers. The degree of malnutrition associated with excessive alcohol intake is another variable. In any case, alcohol may be an additive risk factor for cancer. The recommendation, therefore, is to drink alcohol beverages only in moderation.

An awareness of the recommendations will go a long way towards promoting good health and perhaps preventing cancer.

The awareness should begin at an early age before the onset of cancer induced by environmental factors, and before the lifestyle (especially the dietary habits) is firmly established.

8

General Conclusions

Malnutrition, as a result of either the shortage of food in developing countries or the wrong choice of food in affluent communities, appears to be prevalent in several segments of the world population. The deficiency state is exacerbated when these subjects are treated with drugs, resulting in many cases of overt deficiency with clinical manifestations (Figure 8.1). Such deficiencies may, in turn, result in drug toxicity by impairing the rates of drug metabolism. Thus, the study of the interrelation between nutrients and drugs may be viewed as an extension of the search for factors that modify drug action and dosage. A wide variety of drugs appear to interfere with the availability and utilisation of certain nutrients, and may thus affect the nutritional status of the patient. It may be necessary either to increase the intake of nutrients or to decrease the dosage of a drug in order to counteract any detrimental effect of a drug upon nutritional status, and vice versa. Such considerations are of importance, especially in the risk groups such as the elderly, alcoholics, epileptics, arthritics, pregnant women and infants, where even the normal complex pharmacokinetics of drug metabolism may be further complicated by additional physiological factors.

The level of significance of drug-mediated nutritional deficiency in clinical and sub-clinical situations is, however, often difficult to validate because of variables such as the nutritional status at the onset of therapy, the type of drug involved, the duration of treatment, and age. As a consequence, drug-induced nutrient deficiency is often an overlooked problem. It is important that the probability of this occurrence, as well as the clinical significance of many inter-

Figure 8.1: Consequences of drug-induced nutritional deficiency

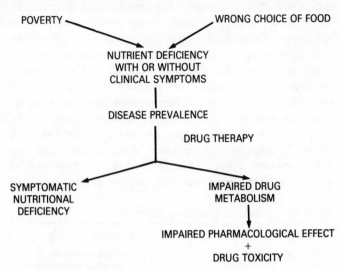

actions, is evaluated in the individual patient rather than by the use of generalised rules.

In view of the current evidence, increasing vigilance on the part of nutritionists and physicians is necessary to avert the potential adverse drug reaction. An outline of investigation procedures in order to identify drug-induced nutrient deficiency is given in Table 8.1. It is important to realise that many clinical signs which are believed to be 'adverse drug reactions' (Table 8.2) may also be signs of nutrient deficiency. Thus, symptoms

Table 8.1: Proposed investigative procedure to identify drug-induced nutrient deficiency

1. Current knowledge of drug-induced nutritional deficiency
2. History of the patient:
    (a) clinical
    (b) drug
    (c) diet
3. Examination of the patient:
    (a) physical
    (b) biochemical
    (c) haematological
    (d) radiological
4. Identification of the drug contributing to a specific nutrient deficiency
5. Confirmation through a follow-up study

such as fatigue and anorexia may be the result of multi-vitamin deficiency. Psychic disturbances may be caused by folacin, B12 or B6 deficiencies, and peptic ulcers may be due to vitamin A deficiency. In addition, there are reactions such as thiazide-diuretic-induced hypokalaemia and anticoagulant-induced haemorrhage which are known to be caused by potassium and vitamin K deficiencies, respectively. Where these clinical symptoms are manifested, supplementation with appropriate nutrients may help to counteract to some extent the adverse drug reactions. In some cases, however, to establish the cause-or-effect relationship between drug intake and nutrient deficiency may require withdrawal of the drug under suspicion, and repeated examinations of the patient at intervals thereafter.

Table 8.2: Some adverse drug reactions

Drug	Reaction
Cardiac glycosides	Fatigue, anorexia, psychic disturbances
Thiazide diuretics	Hypokalaemia, neural dysfunction, arrhythmia
Antihypertensives	Mental depression, dry mouth, diarrhoea
Anticoagulants	Haemorrhage
Hypnotics	Frank psychosis
Tricylic antidepressants	Confusion
Phenylbutazone	Peptic ulcer, hypertension

Careful monitoring of nutritional status is of great importance in all cases where drugs are administered chronically, but particularly so for the elderly (Figure 8.2). The elderly are prone to diseases which are increasingly being treated with drugs. Experimental and clinical evidence suggests that the ability to metabolise and dispose of active drugs decreases progressively with age (O'Malley et al., 1971; Basu, 1980b), primarily because of an impaired hepatic detoxification mechanism and kidney function. Furthermore, the elderly may suffer from deficiencies of various nutrients, either directly, as a result of limited dietary intake, or indirectly, because of drug–nutrient interactions. These factors may account for the known high occurrences of adverse drug reactions in the elderly (Hurwitz, 1969; Greenblatt et al., 1977). Generally speaking, the number of reports of adverse drug reactions received by the

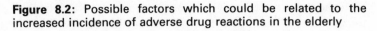

Figure 8.2: Possible factors which could be related to the increased incidence of adverse drug reactions in the elderly

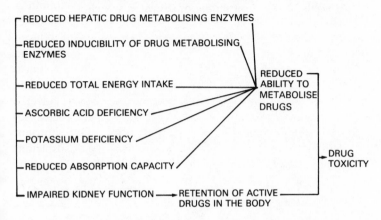

Food and Drug Administration of the United States have increased since 1969: 10 000 reports per year in the 1970s to 40 000 reports in 1985 (Faich *et al.*, 1987). In addition to the conventional pharmacological agents, there has been a growing use of micronutrient supplements in orthomolecular amounts for both preventative and therapeutic purposes. Such uses are justified only in clinical conditions, which are associated with malabsorption syndromes, increased tissue needs, increased excretions, and nutrient-responsive genetic abnormalities. In view of the potential toxicity of vitamins and minerals, any application of these nutrients in doses that are inappropriate or for a purpose that has no scientific basis is cautioned against (Council on Scientific Affairs, 1987).

The human diet may contain a variety of naturally occurring foreign compounds, many of which have been found to be carcinogenic and mutagenic, at least in experimental animals. These compounds are believed to act through the generation of oxygen radicals. The diet also contains a variety of naturally occurring antioxidants as nutrients (i.e. ascorbic acid, α-tocopherol, β-carotene, and selenium), which could be an important aspect of the body's defence mechanism against carcinogens and mutagens. Furthermore, there appears to be a substantial amount of epidemiological and experimental evidence suggesting that various nutrients (other than antioxidants) may be involved in modulating carcinogenesis and

133

mutagenesis. In recent years, prevention of retardation of tumour development by dietary alterations has been one of the major goals of the programme of cancer research.

References

Abel, E.L. (1983). *Marijuana, tobacco, alcohol and reproduction.* CRC Press, Boca Raton, Fla

Abel, E.L. (1985). *Fed. Proc. 44*, 2318

Abel, E.L. and Dintcheff, B.A. (1978). *Pharmacol. Exp. Ther. 207,* 906

Abraham, G.E. and Hargrove, J.T. (1980). *Infertility 3,* 155

Adams, P.W., Wynn, V., Seed, M. and Folkard, J. (1974). *Lancet 2,* 516

Aksoy, M., Basu, T.K., Brient, J. and Dickerson, J.W.T. (1980). *Eur. J. Cancer 16,* 1041

Alam, B.S. and Alam, S.Q. (1985). *Fed. Proc. 44,* 770

Alcoholism and Drug Addictions Research Foundation (1972). *Annual report, 1970. Alcoholism.* ADARF, Toronto

Alhadeff, L., Gualtieri, T. and Lipton, M. (1984). *Nutr. Rev. 42,* 33

Allan, F.N. (1971). In: *Current therapy,* ed. Conn, H.F., W.B. Saunders, Philadelphia, Pa, p. 331

Alleyne, G.A.O. and Young, V.H. (1967). *Clin. Sci. 33,* 189

Alvares, A.P., Anderson, K.E. Conney, A.H. and Kappas, A. (1976). *Proc. Natl Acad. Sci. 73,* 2501

Alvares, A.P., Fischbrin, A., Sassa, S., Anderson, K.E. and Kappas, A. (1976). *Clin. Pharmacol. Ther. 19,* 183

Aly, H.E., Donald, E.A. and Simpson, M.H.W. 1971). *Am. J. Clin. Nutr. 24,* 297

American Institute for Cancer Research (1984). *Planning meals that lower cancer risk: a reference guide.* American Institute for Cancer Research, Washington, DC

Ames, B.N. (1983). *Science 221,* 1256

Anderson, K.E., Conney, A.H. and Kappas, A. (1982). *Nutr. Rev. 40,* 161

Anderson, R. (1981). In: *Vitamin C (ascorbic acid),* eds Counsell, J.N. and Hornig, D.H. Applied Science, London, p. 249

Arakawa, T., Ohara, K. and Kudo, Z. (1973). *Tohoko J. Exp. Med. 110,* 59

Aries,V.C., Crowther, J.S., Drasar, B.S., Hill, M.J. and Williams, R.E.O. (1969). *Gut 10,* 334

Arky, R.A. and Freinkel, N. (1964). *Arch. Intern. Med. 114,* 501

Arky, R.A. and Freinkel, N. (1966). *New Engl. J. Med. 274,* 426

Armstrong, B. and Doll, R. (1975). *Int. J. Cancer 15,* 617

Arvanitakis, C., Chen, G.H., Folscroft, J. and Greenberger, J. (1977). *Gut 18,* 187

Avioli, L.S.W., McDonald, J.E., Lund, J. and DeLuca, H.F. (1967). *J. Clin. Invest. 4,* 983

Azarnoff, D.L., Tucker, D.R. and Barr, G.A. (1965). *Metabolism 14,* 959

Bachrach, W.H. (1959). *Ciba Clin. Symp. 11,* 3

Balmer, S., Howells, G. and Wharton, B.A. (1970). *J. Trop. Pediat. 16*, 20

Bannister, B., Ginsburg, R. and Shneerson, J. (1977). *Br. Med. J. 2*, 738

Baraona, E., Leo, M.A., Borowsky, S.A. and Lieber, C.S. (1977). *J. Clin. Invest.* 60, 546

Baraona, E., Pirola, R.C. and Lieber, C.S. (1973). *J. Clin. Invest. 42*, 296

Barker, B.M. and Bender, D.A. (1980). In: *Vitamins in medicine*, vol. 1, eds Barker, B.M. and Bender, D.A. Heinemann Medical Books, London, p. 348

Barker, I.C. (1957). *Lancet 2*, 747

Barnett, C.A. and Whitney, J.E. (1966). *Metabolism 15*, 88

Bartholomew, C. (1972). *Postgrad. Med. J. 48*, 243

Basu, T.K. (1977a). *J. Hum. Nutr. 31*, 449

Basu, T.K. (1977b). *J. Hum. Nutr. 33*, 24

Basu, T.K. (1977c). *Chem. Biol. Interact. 16*, 247

Basu, T.K. (1979). *Int. J. Vit. Nutr. Res. Suppl.* No. 19, 95

Basu, T.K. (1980a) In: *Clinical implications of drug use*, vol. 1, ed. Basu, T.K. CRC Press, Boca Raton, Fl, p. 1

Basu, T.K. (1980b). In: *Clinical implications of drug use*, vol. 2, ed. Basu, T.K. CRC Press, Boca Raton, Fla, p. 97

Basu, T.K. (1980c). In: *Clinical implications of drug use*, vol. 2, ed. Basu, T.K. CRC Press, Boca Raton, Fla, p. 61

Basu, T.K. (1981). In: *Vitamin C (ascorbic acid)*, eds Counsell, J.N. and Hornig, D.H. Applied Science, London, p. 273

Basu, T.K. (1983a). *Can. J. Physiol. Pharmacol. 61*, 295

Basu, T.K. (1983b). *Can. J. Physiol. Pharmacol. 61*, 1426

Basu, T.K. (1985). *Can. J. Physiol. Pharmacol. 63*, 427

Basu, T.K. (1986). *J. Nutr. 116*, 570

Basu, T.K., Aksoy, M. and Dickerson, J.W.T. (1979). *Chemotherapy 25*, 70

Basu, T.K., Baidoo, S. and Ng, J. (1986). *J. Clin. Biochem. Nutr. 1*, 53

Basu, T.K., Dickerson, J.W.T. (1974). *Chem. Biol. Interact. 8*, 193

Basu, T.K., Dickerson, J.W.T. and Parke, D.V. (1975a). *Nutr. Metab. 18*, 49

Basu, T.K., Dickerson, J.W.T. and Parke, D.V. (1975b). *Nutr. Metab. 18*, 302

Basu, T.K., Raven, R.W., Dickerson, J.W.T. and Williams, D.C. (1974). *Eur. J. Cancer 10*, 507

Basu, T.K. and Schorah, C.J. (1982). *Vitamin C in health and disease.* Croom Helm, London

Basu, T.K., Smethurst, M., Gillett, M.B., Donaldson, D., Jordan, S.J., Williams, D.C. and Hicklin, J.A. (1978). *Acta Vitaminol. Enzymol. 32*, 45

Basu, T.K., Temple, N.J. and Ng, J. (1987). *J. Clin. Biochem. Nutr.* (in press).

Basu, T.K., Weiser, T. and Dempster, J.F. (1984). *Int. J. Vit. Nutr. Res. 54*, 233

Bates, T.R., Sequiera, J.A. and Tembo, A.V. (1974). *Clin. Pharmacol. Ther. 16*, 63

Bauernfeind, J.C. (1980). *The safe use of vitamin A: a report of the International Vitamin A Consultative Group.* Nutrition Foundation, Washington, DC

Bauernfeind, J.C. Newmark, H. and Brin, M. (1974). *Am. J. Clin. Nutr. 27*, 234

Bayliss, E.M., Crowley, J.M., Preece, J.M., Sylvester, P.E. and Marks, V. (1971). *Lancet 1*, 62

Becking, G.C. (1972). *Biochem. Pharmacol. 21*, 1585

Becking, G.C. (1973). *Can. J. Physiol. Pharmacol. 51*, 6

Becking, G.C. (1976). *Fed. Proc. 35*, 2480

Becking, G.C., Morrison, A.B. (1970a). *Biochem. Pharmacol. 19*, 895

Becking, G.C., Morrison, A.B. (1970b). *Biochem. Pharmacol. 19*, 2639

Beckman, C., Roy, R.M. and Sproule, A. 1982). *Mutat. Res. 105*, 73

Beller, G.A., Smith, T.W. and Abelmann, W.H. (1971). *New Engl. J. Med. 284*, 989

Bencze, W.L. (1975). In: *Hypolipidemic agents*, ed. Kritchevsky, D. Springer-Verlag, Berlin and New York, p. 349

Benn, A., Swan, C.H.J., Cooke, W.T., Blair, J.N., Matty, A.J. and Smith, M.E. (1971). *Br. Med. J. 1*, 148

Bergen, S.S. (1974). *Am. J. Dis. Child. 108*, 270

Bernhardt, I.B. and Dorsey, D.J. (1974). *Obstet. Gynecol. 43*, 750

Best, W.R. (1967). *J. Am. Med. Assoc. 201*, 181

Bhattacharyya, M. and DeLuca, H.F. (1974). *Arch. Biochem. Biophys. 160*, 58

Bieri, J.G., Corash, V.S. and Hubbard, N. (1983). *New Engl. J. Med. 308*, 1063

Birt, D.F., Lawson, T.A., Julius, A.D., Runice, C.E. and Salmasi, S. (1982). *Cancer Res. 42*, 4455

Bjelke, E. (1975). *Int. J. Cancer 15*, 561

Blackwell, B., Marley, E. and Mabbit, L.A. (1965). *Lancet 1*, 940

Blundell, J.E., Latham, C.J. and Leshem, M.B. (1976). *J. Pharm. Pharmacol. 28*, 471

Bollag, W. (1974). *Eur. J. Cancer 10*, 731

Bonjour, J.P. (1979). *Int. J. Vit. Nutr. Res. 50*, 96

Bonjour, J.P. (1980a). *Int. J. Vit. Nutr. Res. 50*, 215

Bonjour, J.P. (1980b). *Int. J. Vit. Nutr. Res. 50*, 321

Borchert, P., Mitter, J.A., Miller, E.C. and Shires, T.K. (1973). *Cancer Res. 33*, 590

Bosse, T.R. and Donald, E.A. (1979). *Am. J. Clin. Nutr. 32*, 1015

Boston Collaborative Drug Surveillance Program (1972). *J. Am. Med. Assoc. 220*, 377

Bower, J.O. and Mengle, H.A.K. (1936). *J. Am. Med. Assoc. 106*, 1151

Boyd, E.M., Dobos, I. and Taylor, F.I. (1970). *Chemotherapy 15*, 1

Boyd, E.M. and Taylor, F.I. (1981). *Ind. Med. 38*, 42

Bram, S., Froussard, P., Guichard, M., Jasmin, C., Augery, Y., Sinoussi-Barre, F. and Wray, W. (1980). *Nature (Lond.) 284*, 629

Brater, D.C. and Morrelli, H.F. (1977). *Clin. Pharm. Ther. 22*, 21

Braybrooks, M.P., Barry, B.W. and Abbs, E.T. (1975). *J. Pharm. Pharmacol. 27*, 508

Brenner, A. (1982). *J. Learn. Disabil. 15*, 258

Briggs, M. and Briggs, M. (1973). *Lancet 1*, 998

Briggs, M.H., Gracia-Webb, P. and Davies, P. (1973). *Lancet 2*, 201

Brown, R.R., Rose, D.P., Leklem, J.E., Linkswiler, H. and Anand, R. (1975). *Am. J. Clin. Nutr. 28*, 10

Brubacher, G.B. and Weiser, H. (1985). *Int. J. Vit. Nutr. Res. 55*, 5

Buchler, D. and Warren, J.C. (1966). *Am. J. Obstet. Gynecol. 95*, 479

Bull, L.B., Culvenor, C.J. and Dick, A.T. (1968). *The pyrrolizidine alkaloids*. Elsevier North Holland, Amsterdam

Burkitt, D.P. (1971). *J. Natl. Cancer Inst. 47*, 913

Burkitt, D.P. (1975) *J. Am. Med. Assoc. 231*, 517

Burkitt, D.P., Walker, A.R.P. and Painter, N.S. (1972). *Lancet 2*, 1408

Burton, G.W. and Ingold, K.U. (1984). *Science 224*, 569

Calne, D.B. and Sandler, M. (1970). *Nature (Lond.) 226*, 21

Cameron, E. and Campbell, A. (1974). *Chem. Biol. Interact. 9*, 285

Cameron, E. and Pauling, L. (1978). *Proc. Natl Acad. Sci. 75*, 4538

Campbell, G.D., Steinberg, M.H. and Bower, J.D. (1975). *Ann. Intern. Med. 82*, 810

Campbell, T.C. and Hayes, J.R. (1976). *Fed. Proc. Fed. Am. Soc. Exp. Biol. 35*, 2470

Carlson, L.A., Oro, L. and Ostman, J. (1968). *J. Atheroscler. Res. 8*, 667

Carmichael, B.L. (1973). *Can. J. Hosp. Pharm. 26*, 202

Carpenter, M.P. (1972). *Ann. NY Acad. Sci. 203*, 81

Carpenter, M.P. and Howard, C.N. Jr. (1974). *Am. J. Clin. Nutr. 27*, 966

Carruthers, M.E., Hobbs, C.B. and Warren, R.L. (1966). *J. Clin. Pathol. 19*, 498

Carter, W. and McCarty, K.S. (1966). *Ann. Int. Med. 64*, 1087

Caster, W.O., Wade, A.E., Norred, W.P. and Bargmann, R.E. (1970). *Pharmacology 3*, 177

Catz, C.S., Juchau, M.R. and Yagge, S.J. (1970). *J. Pharmacol. Exp. Ther. 174*, 197

Causins, R.J., DeLuca, H.F. and Gray, R.W. (1970). *Biochemistry 9*, 3649

Cederbaum, A.L., Lieber, C.S. and Rubin, E. (1974). *Arch. Biochem. Biophys. 161*, 26

Cederbaum, A.L., Lieber, C.S., Beattie, D.S. and Rubin, E. (1975). *J. Biol. Chem. 256*, 293

Chanarin, I. (1980). In: *Vitamins in medicine*, vol. 1, eds Barker, B.M. and Bender, D.A. Heinemann, London, p. 247

Chanarin, I., Elmas, P.C. and Mollin, D.L. (1958). *Br. Med. J. 2*, 80

Chanarin, I., Laidlow, J. and Loughridge, L.W. (1960). *Br. Med. J. 1*, 1099

Chin, T.F. and Lach, J.L. (1975). *Am. J. Hosp. Pharm. 32*, 625

Chopra, D.P. and Wilkoff, L.J. (1975). *Proc. Am. Assoc. Cancer Res.* *16*, 35

Christenson, W.N., Ultman, J.E. and Roseman, D.M. (1957). *J. Am. Med. Assoc. 163*, 940

Chu, E.W. and Malmgren, R.A. (1965). *Cancer Res. 25*, 884

Chung, E.K. (1970). *Am. Heart J. 79*, 845

Ch'ien, L.T., Krundieck, C.L., Scott, C.W. and Butterworth, C.E. (1975). *Am. J. Clin. Nutr. 28*, 51

Clark, L.C. (1985). *Fed. Proc. 44*, 2584

Clemetson, C.A.B. (1968). *Lancet 2*, 1037

Cochrane, W. (1965). *Can. Med. Assoc. J. 83*, 893

Cohen, S.M., Wittenberg, J.F. and Bryan, G.T. (1976). *Cancer Res. 36*, 2334

Colby, H.D., Kramer, R.E., Greiner, J.W., Robinson, D.A., Krause, R.F. and Canady, W.J. (1975). *Biochem. Pharmacol. 24*, 1644

Combs, G.F. and Clark, L.C. (1985). *Nutr. Rev. 43*, 325

Committee on Diet, Nutrition, and Cancer (1982). In: *Diet, nutrition and cancer*. National Academy Press, Washington, DC

Committee on Nutrition, Academy of Pediatrics (1963). *Pediatrics 31*, 512

Concon, J.M., Newburg, D.S. and Swerczek, T.W. (1979). *Nutr. Cancer 1*, 22

Cone, M.V. and Nettesheim, P. (1973). *J. Natl Cancer Inst. 50*, 1599

Conney, A.H. (1967). *Pharmacol. Rev. 19*, 317

Conney, A.H., Bray, G.A., Evans, C. and Burns, J.J. (1961). *Ann. NY Acad. Sci. 92*, 115

Conney, A.H., Pantuck, E.J., Hsiao, K.C., Garland, W.A., Anderson, K.E., Alvares, A.P. and Kappas, A. (1976). *Clin. Pharmacol. Ther. 20*, 633

Cook, J.W., Kennaway, E.L. and Kennaway, N.M. (1940). *Nature (Lond.) 145*, 627

Cook, M.G. and McNamara, P. (1980). *Cancer Res.* 40, 1329

Coronary Drug Project Research Group (1972). *J. Am. Med. Assoc. 220*, 996

Coronary Drug Project Research Group (1975). *J. Am. Med. Assoc. 231*, 360

Coronato, A. and Glass, G.B.J. (1973). *Proc. Soc. Exp. Biol. Med. 142*, 1341

Corrigan, J.J., and Marcus, F.I. (1974). *J. Am. Med. Assoc. 230*, 1300

Corstens, F., Kerremans, A. and Claessens, R. (1986). *J. Nucl. Med. 27*, 219

Cotzias, G.C. (1969). *J. Am. Med. Assoc. 210*, 1255

Council on Scientific Affairs (1987). *J. Am. Med. Assoc. 257*, 1929

Cowman, D.H. (1973). *Lab. Clin. Med. 81*, 64

Craft, A.W., Kay, A.G.M., Lawson, D.N. and McElwain, T.J. (1977). *Br. Med. J. 2*, 1511

Craft, I.L. and Peters, T.J. (1971). *Clin. Sci. 41*, 301

Creagen, E.T., Moertel, C.G., O'Fallen, J.R., Schutt, A.J., O'Connell, M.J., Rubin, J. and Frytak, S. (1979). *New Engl. J. Med. 301*, 687

Crocker, T.T. and Sanders, L.L. (1970). *Cancer Res. 30*, 1312

Crounse, R.G. (1961). *J. Invest. Dermat. 37*, 529

Dalen N. and Lamke, B. (1976). *Acta Orthop. Scand. 47*, 469

Daliam, A., Savoure, N., Cottencin, C. and Nicol, M. (1986). *Bull. Cancer (Paris) 73*, 243

Darby, W.J. (1979). In: *Fermented food beverages in nutrition*, eds Gastineau, C.F., Darby, W.J. and Turner, T.B. Academic Press, New York, p. 61

D'Arcy, P.F. and Griffin, J.P. (1972). *Iatrogenic diseases*. Oxford University Press, London

Datey, K.K., Kelar, P.N. and Pandya, R.S. (1973). *Br. J. Clin. Pract. 27*, 373

Davenport, H.W. (1977). *Physiology of the digestive tract*, 4th edn. Yearbook Med. Pub., Chicago, Ill., p. 194

Davidson, W. (1971). *Br. J. Hosp. Med. 6*, 83

Davies, L., Hastrop, K. and Bender, A.E. (1973). *Mod. Geriatr. 3*, 482

Davies, R.E. (1967). *Cancer Res. 27*, 237

Davis, R.E. and Smith, B.K. (1974). *Med. J. Aust. 2*, 357

DeCosse, J.J., Adams, M.B., Kuzma, J.F., Logerfo, P. and Condon, R.E. (1975). *Surgery 78*, 608

DeLuca, H.F. (1974). *Fed. Proc. 33*, 2211

DeLuca, L., Maestri, N., Bonnani, F. and Nelson, D. (1972). *Cancer 30*, 1326

Dent, C.E., Richens, A., Rowe, D.J.F. and Stamp, T.C.B. (1970). *Br. Med. J. 4*, 69

DeWulf, H., Stalmans, W. and Hers, H.G. (1970). *Eur. J. Biochem. 15*, 1

Dickerson, J.W.T. (1978). In: *Nutrition and the clinical management of disease*, eds Dickerson, J.W.T. and Lee, H.A., Chapter 14. Edward Arnold, London

Dickerson, J.W.T., Basu, T.K. and Parke, D.V. (1976). *J. Nutr. 106*, 258

Dingell, J.V., Joiner, P.D. and Hurwitz, L. (1966). *Biochem. Pharmacol. 15*, 971

Dingle, J.T. and Lucy, J.A. (1965). *Biol. Rev. 40*, 422

Dion, L.D., Blalock, J.F. and Gifford, G.E. (1978). *Exp. Cell. Res. 117*, 15

Dipalma, J.R. and Ritchie, D.M. (1977). *Ann. Rev. Pharm. Toxicol. 17*, 133

Dobbins, W.O. (1968). *Gastroenterology 54*, 1193

Donald, E.A. and Bosse, T.R. (1979). *Am. J. Clin. Nutr. 32*, 1024

Drash, A., Elliott, J., Langs, H., Lavenstein, A.F. and Cooke, R.E. (1966). *Clin. Pharmacol. Ther. 7*, 340

Druskin, M.S., Wallen, M.H. and Bonagura, L. (1962). *New Engl. J. Med. 267*, 483

Dunbow, M.H. and Burchell, H.B. (1965). *Ann. Intern. Med. 62*, 956

Edgar, J.A. (1970). *Nature (Lond.) 227*, 24

Eichner, E.R. and Hillman, R.S. (1971). *Am. J. Med. 50*, 218

Eichner, E.R. and Hillman, R.S. (1973). *J. Clin. Invest. 52*, 584

Eichner, E.R., Pierce, H.I. and Hillman, R.S. (1971). *New Engl. J. Med. 284*, 933

Einstein, N., Baker, A., Galper, J. and Wolfe, H. (1975). *Am. J. Dig. Dis. 20*, 282

Eisalo, A., Ahrenberg, P. and Nikkila, E.A. (1963). *Acta Med. Scand. 173*, 639

Eisenstein, A. (1973). *Am. J. Clin. Nutr. 26*, 113

Eisenstein, A., Spencer, S., Flatness, S. and Brodsky, A. (1966). *Endocrinology 79*, 182

Elgee, N.J. (1970). *Ann. Intern. Med. 72*, 409

Ellis, J.M., Folkers, K., Levy, M., Shikykuishi, S., Lewandowski, J., Nishii, S., Schubert, H.A. and Ulrich, R. (1982). *Proc. Natl Acad. Sci. 79*, 7494

Embil, K., Litwiller, D.C., Lepore, R.A., Field, F.P. and Torosian, G. (1976). *Am. J. Hosp. Pharm. 33*, 1294

Endo, H. and Takahashi, K. (1973). *Nature (Lond.) 245*, 325

Enloe, C.F. (1980). *Nutrition Today* Sept/Oct., p. 14

Ernst, V., Levin, D.H. and London, I.M. (1978). *Proc. Natl Acad. Sci. USA 75*, 4110

Evans, D.A.P., Manley, K.A. and McKusick, V.A. (1960). *Br. Med. J. 2*, 485

Evans, C.D.H. and Lacey, J.H. (1986). *Br. Med. J. 292*, 509

Faich, G.A., Knapp, D., Dreis, M. and Turner, W. (1987). *J. Am. Med. Assoc. 257*, 2068

Fairweather, F.A. (1981). *Proc. Nutr. Soc. 40*, 21

Faizy, A. and Mahatab, B.S. (1976). *Contraception 14*, 309

Faizy, A., Bamji, M.S. and Iyengar, L. (1975). *Am. J. Clin. Nutr. 28*, 606

Fajans, S.S., Floyd, J.C., Knopf, R.F., Rull, J., Guntsche, E.M. and Conn, J.W. (1966). *J. Clin. Invest. 45*, 481

Faller, J. and Fox, I.H. (1982). *New Engl. J. Med. 307*, 1598

Faloon, W.W. (1966). *Ann. NY Acad. Sci. 132*, 879

Faloon, W.W. (1970). *Am. J. Clin. Nutr. 23*, 645

Fehily, A.M., Phillops, K.M. and Yarnell, J.W.G. (1980) *Am. J. Clin. Nutr. 40*, 827

Fiddler, W., Pensabene, J.W., Piotrowski, E.G., Doerr, R.C. and Wasserman, A.E. (1973). *J. Fd Sci. 38*, 1084

Field, J.B. and Mandell, S. (1964). *Metabolism 13*, 959

Field, J.B., Williams, H.E. and Mortimere, G.E. (1963). *J. Clin. Invest. 42*, 497

Fingl, E. (1975). In: *The pharmacological basis of therapeutics*, eds Goodman, L.S. and Gilman, A.Z. Macmillan, New York, p. 978

Fischer, L.J., Millburn, P., Smith, R.L. and Williams, R.T. (1966). *Biochem. J. 100*, 69P

Flohe, L. (1982). In: *Free radicals in biology*, vol. 5, ed. Pryor, W.A., Academic Press, New York, p. 223

Foods and Nutrition Board, National Research Council. (1980) *Recommended Dietary Allowances*, 9th edn. National Academy of Sciences, Washington, DC

Foote, C.S. (1982). In: *Pathology of oxygen*, ed. Autor, A. Academic Press, New York, p. 21

Fraser, D.R. and Kodicek, E. (1970). *Nature (Lond.) 228*, 764

Fraumene, J.F. (1967). *J. Am. Med. Assoc. 201*, 150

Freinkel, N., Singer, D.L., Arky, R.A., Bleicher, S.J., Anderson, J.B. and Silbert, C.K. (1963). *J. Clin. Invest. 42*, 1112

Frick, P.G., Hitzig, W.H. and Betke, K. (1962). *Blood 20*, 261

Frick, P.G., Riedler, G. and Brogli, H. (1967). *J. Appl. Physiol. 23*, 387

Fuld, H. and Moorhouse, E.H. (1956). *Br. Med. J. 1*, 1021

Furlanut, M., Benetello, P., Avogaro, A. and Dainese, R. (1978). *Clin. Pharmacol. Ther. 24*, 294

Gal, I., Sharman, I.M. and Prys-Davies, J. (1972). *Adv. Teratol. 5*, 143

Gastineau, C.F., Darby, W.J. and Turner, T.B. (1979). *Fermented food beverages in nutrition*, Academic Press, New York, p. 206

Genta, V.M., Kaufman, D.G., Harris, C.C., Smith, J.M., Sporn, M.B. and Saffiotti, V. (1974). *Nature (Lond.) 247*, 48

Gershberg, H., Javier, Z. and Hulse, M. (1964). *Diabetes 13*, 378

Gerson, C.D. (1971). *Gastroenterology 63*, 246

Gey, K.F. and Pletscher, A. (1964). *Biochem. J. 92*, 300

Gibbard, F.B., Nicholls, A. and Wright, M.G. (1981). *Eur. J. Clin. Pharmacol. 19*, 57

Gill, J.H., Zezulke, A.V., Shipley, M.J., Gill, S.K. and Beevers, D.G. (1986). *New Engl. J. Med. 315*, 1041

Gillespie, N.G., Mena, I., Cotzias, G.C. and Bell, M.A. (1973). *J. Am. Diet. Assoc. 62*, 525

Ginsberg, H., Olefsky, J., Farquhar, J.W. and Reaven, G.M. (1974). *Ann. Intern. Med. 80*, 143

Ginter, E. (1977). *Int. J. Vit. Nutr. Res. Suppl.* No. 16, p. 53

Ginter, E., Kosinova, A., Hudecova, A. and Mlynarcikova, V. (1984). *J. Nutr. 114*, 485

Ginter, E., Vejmolova, J. (1981). *Br. J. Clin. Pharmacol. 12*, 256

Giroud, A. (1970). *The nutrition of the embryo*. Charles C. Thomas, Springfield, Ill.

Goldin, B.R. and Gorbach, S.L. (1976). *J. Natl Cancer Inst. 57*, 371

Goldstein, A. (1949). *Pharmacol. Rev. 1*, 102

Graham, S. (1984). *J. Natl Cancer Inst. 73*, 1423

Granenus, A.K., Jagenburg, R., Rodjer, S. and Svanborg, A. (1971). *Br. Med. J. 4*, 262

Grantham, P.H., Horton, R.E., Weisburger, E.K. and Weisburger, J.H. (1970). Biochem. Pharmacol. 19, 163

Gray, G.M. (1973). *Am. J. Clin. Nutr. 26*, 121

Greenberger, N.J. (1973). *Am. J. Clin. Nutr. 26*, 104

Greenblatt, D.J., Allen, M.D. and Shader, K.I. (1977). *Clin. Pharmacol. Ther. 21*, 355

Gregor, A., Lee, P.N., Roe, J.C., Wilson, M.J. and Melton, A. (1980). *Nutr. Cancer 2*, 93

Griffith, P.R. and Innes, F.C. (1983). *Nutr. Res. 3*, 445

Gunn, A.D.G. (1985). *Int. J. Vit. Nutr. Res. Suppl.* 27, *213*

Guttemplan, J.B. (1977). *Nature (Lond.) 268*, 368

Gydell, K. (1957). *Acta Haematol. 17*, 1

Haddow, A. (1958). *Br. Med. Bull. 14*, 79

Haenszel, W., Berg, J.W., Segi, M., Kurihara, M. and Locke, F.B. (1973). *J. Natl Cancer Inst. 51*, 1765

Haenszel, W. and Correa, P. (1971). *Cancer 28*, 14

Haenszel, W., Correa, P. and Cuello, C. (1975). *J. Natl Cancer Inst. 54*, 1031

Haenszel, W., Kurihara, M., Segi, M. and Lee, R.K.C. (1972). *J. Natl Cancer. Inst. 49*, 969

Hahn, T.J. and Alvioli, L.V. (1975) *Arch. Intern. Med. 135*, 997

Hallstrom, T. (1969). *Acta Psychiat. Scand. 45*, 19

Hanck, A. and Weiser, H. (1977). *Int. J. Vit. Nutr. Res. Suppl.* No. 16, 67

Hara, T. and Taniguchi, M. (1982). *Biochem. Biophys. Res. Commun. 104*, 394

Harbison, R.D. and Becker, B.A. (1972). *Toxicol. Appl. Pharmacol. 22*, 193

Haroz, R.K. and Thomasson, J. (1980). *Toxicol. Lett. Suppl. 6*, 72

Harris, R.W.C., Forman, D., Doll, R., Vessey, M.P., Wald, N.J. (1986). *Br. J. Cancer 53*, 653

Harrison, Y.E. and West, W.L. (1971). *Biochem. Pharmacol. 20*, 2105

Hartshorn, E.A. (1977). *J. Am. Diet. Assoc. 70*, 15

Harttroft, S.W. and Porta, E.A. (1968). *Can. J. Physiol. Pharmacol. 46*, 463

Harvengt, C., De Schepper, P., Lamay, F. and Hansen, J. (1973). *J. Clin. Pharmacol. 13*, 36

Hashim, S.A., Bergan, S.S. Jr and Van Itallie, T.B. (1961). *Proc. Soc. Exp. Biol. Med., 106*, 173

Hashim, S.A. and Van Itallie, T.B. (1965). *J. Am. Med. Assoc. 192*, 289

Hauswirth, J.W. and Nair, P.O. (1975). *Am. J. Clin. Nutr. 28*, 1087

Havel, R.J. and Kane, J.P. (1973). *Ann. Rev. Pharmacol. 13*, 287

Havlicek, V. and Childaeva, R. (1976). *Lancet 2*, 477

Hawkins, C.F. and Meynell, M.J. (1958). *Quart. J. Med. 27*, 45

Hecht, A. and Goldner, M.G. (1959). *Metabolism 8*, 418

Hell, D., Six, P. and Salkeld, R. (1976). *Schweiz. Med. Wschr. 106*, 1466

Hennekens, C.H., Mayrent, S.L. and Willett, W. (1986). *Cancer 58*, 1837

Herbert, V. and Jacob, E. (1974). *J. Am. Med. Assoc. 230*, 241

Herbst, A.L., Ulfelder, H. and Poskanzer, D.C. (1971). *New Engl. J. Med. 284*, 878

Herzberg, B.N., Johnson, A.L. and Brown, S. (1970). *Br. Med. J. 4*, 142

143

REFERENCES

Hietanen, E., Laitinen, M., Vainio, H. and Hanninen, O. (1975). *Lipids 10*, 467

Hill, D.L. and Grubbs, C.J. (1982). *Anticancer Res. 2*, 111

Hill, D.L. and Shih, T.W. (1974). *Cancer Res. 34*, 564

Hill, M.J. (1971). In: *Some implications of steroid hormones in cancer*, eds Williams, D.C. and Briggs, M.H. Heinemann, London, p. 94

Hill, M.J. (1974). *Am. J. Clin. Nutr. 27*, 1475

Hill, M.J., Draser, B.S., Aries, V.C., Crowther, J.S., Hawksworth, G. and Williams, R.E.O. (1971). *Lancet 1*, 95

Hill, M.J., Draser, B.S., Williams, R.E.O., Meade, T.W., Cox, A.G., Simpson, J.E.P. and Morson, B.C. (1975). *Lancet 2*, 535

Hirayama, T. (1979). *Nutr. Cancer 1*, 67

Hodge, J.V., Nye, E.R. and Emerson, G.W. (1964). *Lancet 1*, 1108

Hoefel, O.S. (1977). *Int. J. Vit. Nutr. Res. Suppl.* No. 16, 127

Hoensch, H., Woo, C.H. and Schmid, R. (1975). *Biochem. Biophys. Res. Commun. 65*, 399

Hoffbrand, A.V. (1971). *J. Clin. Pathol.* (S24) 5, 66

Hoffer, A. (1977). In: *A physician's handbook on orthomolecular medicine*, eds Williams, R.J. and Kalita, D.K. Keats, Connecticut, p. 83

Hogenkamp, H.P.C. (1980). *Am. J. Clin. Nutr. 33*, 1

Hommes, O.R. and Obbens, E.A.M.T. (1972). *J. Neurol. Sci. 16*, 271

Hornig, D., Weiser, H., Weber, F. and Wiss, O. (1973). *Int. J. Vit. Res. 43*, 28

Hornykiewicz, O. (1966). *Pharmacol. Rev. 18*, 925

Houben, P.F.M., Hommes, O.R. and Knaven, P.J.H. (1971). *Epilepsia 12*, 235

Hourihave, D.O'B. and Weir, D.G. (1970). *Br. Med. J. 1*, 86

Hughes, M.R., Baylink, D.J., Jones, P.G. and Haussler, M.R. (1976). *J. Clin. Invest. 58*, 61

Hurwitz, A. and Sheehan, M.B. (1971). *J. Pharmacol. Exp. Ther. 179*, 124

Hurwitz, N. (1969). *Br. Med. J. 1*, 549

Hutchinson, T.A., Polansky, S.M. and Feinstein, A.R. (1979). *Lancet 2*, 705

IARC (1972). *Monograph on the evaluation of the carcinogenic risk of chemicals to man: Natural Products*, vol. 1. International Agency for Research on Cancer, Lyon, p. 145

Iber, F.L. (1977). *Clin. Pharmacol. Ther. 22*, 735

Ikonopisov, R.L. (1972). In: *Melanoma and skin cancer*, ed. McCarthy, W.H. International Union Against Cancer, Sydney, p. 223

Ioannides, C. and Parke, D.V. (1973). *Biochem. Soc. Trans. 1*, 716

Ioannides, C. and Parke, D.V. (1979). *J. Hum. Nutr. 33*, 357

Ioannides, C., Stone, A.N., Breacker, P.J. and Basu, T.K. (1982). *Biochem. Pharmacol. 31*, 4035

Isbell, H. and Chrusciel, T.L. (1970). *WHO Bull. Suppl. 43*, 66

Issenberg, P. (1976). *Fed. Proc. 35*, 1322

Ivey, M. (1979). *Handbook of non-prescribed drugs*, 6th edn, American Pharmaceutical Association, Washington, DC

Jacob, H.S. and Winterhalter, K.H. (1970). *J. Clin. Invest. 49*, 2008

Jacob, H.S. (1970). *Sem. Hematol. 7*, 341

Jacobs, M.M. (1983). *Cancer Res. 43*, 1646

Jacobson, E.D., Chodos, R.B. and Faloon, W.W. (1960). *Am. J. Med. 28*, 524

Jaffe, J.A. (1969). *Ann. NY Acad. Sci. 166*, 57

Jaffe, J.A., Colaizzi, J.L. and Barry, H. (1971). *J. Pharm. Sci. 60*, 1646

Janz, D. (1975). *Epilepsia 16*, 159

Jarvell, K.F., Christoffersen, T. and Morland, J. (1965). *Arch. Biochem. Biophys. 111*, 15

Jaya Rao, K.S. (1974). *Lancet 1*, 709

Jensen, O.N. and Olesen, O.V. (1969). *Arch. Neurol. 21*, 208

Jepson, E.M. (1963). *Br. Med. J. 1*, 1446

Johnell, O., Nilsson, B.E. and Wiklund, P.E. (1982). *Clin. Orthop. 165*, 253

Jones, W.A. and Jones, G.P. (1953). *Lancet 1*, 1073

Joshi, V.G., Eswaran, S., Nagesh, R.,, Pai, M., Mathewa, G. and Mallick, P.N. (1982). *J. Orthomol. Psychiat. 11*, 45

Kakar, S.C. and Wilson, C.W.M. (1974). *Proc. Nutr. Soc. 33*, 110A

Kakar, S.C. and Wilson, C.W.M. (1976). *Proc. Nutr. Soc. 35*, 9A

Kallner, A. (1981). In: *Vitamin C (ascorbic acid),* eds Counsell, J.N. and Hornig, D.H., Applied Science, London, p. 63

Kallner, A., Hartmann, D. and Hornig, D.H. (1981). *Am. J. Clin. Nutr. 34*, 1347

Kallstrom, B. and Nylof, R. (1969). *Acta Psychiat. Scand. 45*, 137

Kalter, H. and Warkany, J. (1961). *Am. J. Pathol. 38*, 1

Kamm, J.J., Dashman, T., Conney, A.H. and Burns, J.J. (1973). *Proc. Natl Acad. Sci. 70*, 747

Kansel, P.C., Buse, J. and Buse, M.G. (1969). *Southern Med. J. 62*, 1374

Kappas, A., Anderson, K.E., Conney, A.H. and Alvers, A.P. (1976). *Clin. Pharmacol. Ther. 20*, 643

Kato, R. (1967). *Biochem. Pharmacol. 16*, 871

Kato, R., Oshima, T. and Tomizawa, S. (1968). *Jap. J. Pharmacol. 18*, 356

Kent, J.C., Devlin, R.D., Gutteridge, D.H. and Retallack, R.W. (1979). *Biochem. Biophys. Res. Commun. 89*, 155

Kent, S.C. and Durack, D.T. (1978). *Ann. Intern. Med. 88*, 520

Kishi, H., Kishi, T., Williams, R.H., Watanabe, T., Folkers, K. and Stahl, M.L. (1977). *Res. Commun. Chem. Pathol. Pharmacol. 17*, 283

Klipstein, F.A. (1964). *Blood 23*, 68

Kniffen, J.C., Noyas, W.D. and Porter, F.S. (1970). *Clin. Res. 18*, 38

Kolata, G.B. (1975). *Science 187*, 635

Korsan-Bengsten, K., Elmfeldt, D. and Holm, T. (1974). *Thromb. Diath. Haemorrh. 31*, 505

Kosower, N.S., Vanderhoff, G.A. and Kosower, E.M. (1972). *Biochim. Biophys. Acta 272*, 623

145

Krasner, N. and Dymock, I.W. (1974). *Br. J. Cancer 30*, 14

Krattila, K. and Kangas, L. (1977). *Acta Pharmacol. Toxicol. 40*, 241

Krawitt, E.L., Grundman, M.J. and Mawer, B. (1977). *Lancet 2*, 1246

Krebs, H.A., Freedland, R.A., Hems, R. and Stubbs, M. (1969). *Biochem. J. 112*, 117

Krinsky, N.I. and Deneke, S.M. (1982). *J. Natl Cancer Inst. 69*, 205

Kritchevsky, D. (1960). *Metabolism 9*, 984

Kummet, T. and Meyskens, F.L. (1983). *Sem. Oncol. 10*, 281

Kuntzman, R., Pantuck, E.J., Kaplan, S.A. and Conney, A.H. (1977). *Clin. Pharmacol. Ther. 22*, 757

Kurek, M.P. and Corwin, L.M. (1982). *Nutr. Cancer 4*, 128

Kyrtopoulos, S.A. (1987). *Am. J. Clin. Nutr. 45*, 1344

Lambert, B., Brisson, G. and Bielmann, P. (1981). *Gynecol. Oncol. 11*, 136

Lambert, L. and Wills, E.D. (1977). *Biochem. Pharmacol. 26*, 1423

Lambert, M.L. (1975). *Am. J. Nutr. 75*, 402

Lamden, M.P. (1971). *New Engl. J. Med. 284*, 336

Lancet Editorial. (1983). *Lancet 1*, 682

Lang, M. (1976). *Gen. Pharmacol. 7*, 415

Langer, T., Levy, R.I. and Fredrickson, D.S. (1969). *Circulation 40*, 111

Langer, T., Strober, W. and Levy, R.I. (1972). *J. Clin. Invest. 51*, 1528

Laqueur, G.L. and Spatz, M. (1968). *Cancer Res. 28*, 2262

Lasnitzki, I. (1976). *Br. J. Cancer 34*, 239

Latham, A.N., Millbank, L. and Richens, A. (1973). *J. Clin. Pharmacol. 13*, 337

Lawson, D.E.M., Fraser, D.R., Kodicek, E., Morris, H.R. and Williams, D.H. (1971). *Nature (Lond.) 230*, 228

Leboeuf, R.A. and Hoekstra, W.G. (1983). *J. Nutr. 113*, 845

Leboeuf, R.A. and Hoekstra, W.G. (1985). *Fed. Proc. 44*, 2563

Lecocq, F.R., Mebane, D. and Madison, L.L. (1964). *J. Clin. Invest. 43*, 237

Leevy, C.M., Baker, H., Ten Hove, W., Frank, O. and Gillene, R. (1965). *Am. J. Clin. Nutr. 16*, 339

Lefevre, A.F., Decarli, L.M. and Lieber, C.S. (1972). *J. Lipid Res. 13*, 48

Lehmann, P. (1978). *Food and drug interactions,* FDA Consumer, New Publication, No. (FDA) 78-3070, US Govt Printing Office, Washington, DC

Leklem, J.E., Brown, R.R., Rose, D.P., Linkswiller, H. and Arend, R.A. (1975). *Am. J. Clin. Nutr. 28*, 535

Leo, M.A. and Lieber, C.S. (1982). *New Engl. J. Med. 307*, 597

Leo, M.A., Sato, M. and Lieber, C.S. (1981). *Clin. Res. 29*, 266

Leon, A.S., Spiegel, H.E., Thomas, G. and Abrams, W.B. (1971). *J. Am. Med. Assoc. 218*, 1924

Levine, N. and Meyskens, F.L. (1980). *Lancet 2*, 224

Levine, R.L. (1970). *Dig. Dis. 15*, 171

Levy, G. and Jusko, W. (1965). *J. Pharm. Sci. 54*, 219

Lieber, C.S. (1967). *Ann. Rev. Med. 18*, 35
Lieber, C.S. (1973). *New Engl. J. Med. 288*, 356
Lieber, C.S. (1978). *New Engl. J. Med. 298*, 888
Lieber, C.S. (1984). *New Engl. J. Med. 310*, 846
Lieber, C.S. and DeCarli, L.M. (1970). *Life Sci. 9*, 267
Lieber, C.S. and DeCarli, L.M. (1974). *J. Med. Primatol. 3*, 153
Lieber, C.S. and DeCarli, L.M. (1975). *Proc. Natl Acad. Sci. 72*, 437
Lieber, C.S. and DeCarli, L.M. (1977). In: *Metabolic aspects of alcoholism*, Lieber, C.S. ed. MTP Press, Lancaster, p. 45
Liedholm, H., Wahlin-Boll, Hanson, A. and Melander, A. (1982). *Drug–Nutr. Interact. 1*, 293
Lijinsky, W. (1974). *Cancer Res. 34*, 255
Lijinsky, W. and Epstein, S.S. (1970). *Nature (Lond.) 225*, 21
Little, R.E. (1977). *Am. J. Public Hlth 67*, 1154
Logue, T. and Frommer, D. (1980). *Australia and New Zealand J. Med. 10*, 588
Long, E.A., Nelson, A.A., Fitzhigh, O. and Hansen, W.H. (1963). *Arch. Pathol. 75*, 595
Long, R.G., Wills, M.R., Skinner, R.K. and Sherlock, S. (1976). *Lancet 2*, 650
Lotan, R. (1980). *Biochim. Biophys. Acta 605*, 33
Lown, B. (1956). *Adv. Intern. Med. 8*, 125
Lucarotti, R.L., Barry, H. and Poust, R.I. (1972). *J. Pharm. Sci. 61*, 903
Luft, D., Degkwitz, E., Hochli-Kaufman, L. and Staudinger, H. (1972). *Hoppe-Seyler's Z. Physiol. Chem. 353*, 1420
Luhby, A.L., Brin, M., Gordon, M., David, P., Murphy, M. and Spiegel, H. (1971). *Am. J. Clin. Nutr. 24*, 684
Lumeng, L., Cleary, R.E. and Li, T.K. (1974). *Am. J. Clin. Nutr. 27*, 326

Macdonald, H., Place, V.A., Falk, H. and Darken, M.A. (1967). *Chemotherapia 12*, 282
MacIntyre, I. (1979). *Adv. Nephrol. 8*, 153
Macmillan, D.C., Oliver, M.F. and Simpson, J.D. (1965). *Lancet 2*, 924
Madappally, M.M., Mackerer, C.R. and Mehlman, M.A. (1972). *Life Sci. 11*, 77
Magee, P.N. and Lee, K.Y. (1964). *Biochem. J. 91*, 35
Maier, B.R., Flynn, M.A., Burton, G.C., Tsutakawa, P.K. and Hentages, D.J. (1974). *Am. J. Clin. Nutr. 27*, 1470
Majunder, S.K., Shaw, G.K. and Thomson, A.D. (1983). *Int. J. Vit. Nutr. Res. 53*, 273
Makela, K., Room, R., Single, E., Sulkunen, P., Walsh, B., Bunce, R., Cahannes, M. *et al.* (1981). *Alcohol, Society and the State*. Addiction Res. Found. Toronto, Ontario
Makiura, S., Aoe, H., Sugihara, S., Hirao, K., Masayuki, A. and Ito, N. (1974). *J. Natl Cancer Inst. 53*, 1253
Malcolm, A.D., Mae, P.M., Outar, K.P. and Pawan, G.L.S. (1972). *Proc. Nutr. Soc. 31*, 12A

Mao, C.C. and Jacobson, E.D. (1970). *Am. J. Clin. Nutr. 23*, 820

Maragoudakis, M.E. (1970). *Biochemistry (Wash.), 9*, 413

Marks, V. (1974). *Proc. Nutr. Soc. 33*, 209

Mars, H. (1974). *Arch Neurol. 30*, 444

Marshall, J.W. and McLean, A.E.M. (1969). *Biochem. Pharmacol. 18*, 153

Marshall, J.W. and McLean, A.E.M. (1971). *Biochem. J. 122*, 569

Mason, D.T. (1974). *Ann. Intern. Med. 80*, 520

Mason, R.P. and Holtzman, J.L. (1975). *Biochemistry 14*, 1626

Mathews-Roth, M.M. (1982). *Oncology 39*, 33

Mathews-Roth, M.M. (1985). *Pure Appl. Chem. 57*, 717

Matsunage, F., Kubo, A. and Katakura, G. (1963). *J. Ther. (Tokyo) 45*, 1988

Mattson, R.H., Gallagher, B.B., Reynolds, E.H. and Glass, D. (1973). *Arch. Neurol. 29*, 78

Maugiere, F., Quoex, C. and Bello, S. (1975). *Epilepsia 16*, 535

Maxwell, J.D., Hunter, J., Stewart, D.A., Ardeman, S. and Williams, R. (1972). *Br. Med. J. 1*, 297

McCalla, D.R., Reuvers, A. and Kaiser, C. (1970). *J. Bacteriol. 104*, 1126

McCarthy, C.G. and Finland, M. (1960). *New Engl. J. Med. 263*, 315

McClain, C.J., Van Thiel, D.H., Parker, S., Badzin, L.K. and Gilbert, H. (1979). *Alcohol Clin. Exp. Res. 3*, 135

McCormick, D.L., Burns, F.J. and Albert, R.E. (1980). *Cancer Res. 40*, 1140

McCraken, G.H. Jr., Grinsburg, C.M., Clahsen, J.C. and Thomas, M.L. (1978). *Pediatrics 62*, 738

McLaren, D.S. (1964). *Nutr. Rev. 22*, 289

McLean, A.E.M. and McLean, E.K. (1966). *Biochem. J. 100*, 564

McLean, A.E.M. and Verschuuren, H.G. (1969). *Br. J. Exp. Pathol. 50*, 22

McLean, A.J., Barron, K., du Souich, P., Haegele, K.D., McNay, J.L., Carrier, O. and Briggs, A. (1978). *J. Pharmacol. Exp. Ther. 205*, 418

McLeroy, V.J. and Schendel, H.E. (1973). *Am. J. Clin. Nutr. 26*, 191

Mergens, W.J., Vane, F.M., Tannenbaum, S.R., Green, L. and Skipper, P.L. (1979). *J. Pharm. Sci. 68*, 827

Mehta, S., Kalsi, H.K., Jayaraman, S. and Mathur, V.S. (1975). *Am. J. Clin. Nutr. 28*, 977

Melander, A. (1978). *Clin. Pharmacokinet. 3*, 337

Melander, A., Danielson, K., Hanson, A., Rydell, B., Schersten, B., Thulin, T. and Wahlin, E. (1977a). *Clin. Pharmacol. Ther. 22*, 104

Melander, A., Danielson, K., Schersten, B. and Wahlin, E. (1977b). *Clin. Pharmacol. Ther. 22*, 108

Mena, I. and Cotzias, G.C. (1975). *N. Engl. J. Med. 292*, 181

Mengel, C.E. and Green, H.L. (1976). *Ann. Intern. Med. 84*, 490

Menkes, M.S., Comstock, G.W., Vuilleumier, J.P., Helsing, K.J., Rider, A.A. and Brookmeyer, R. *N. Engl. J. Med. 315*, 1250

Mettlin, C., Graham, S. and Swanson, M. (1979). *J. Natl Cancer Inst. 62*, 1435

Meyer, L.M., Heeve, W.L. and Bertscher, R.W. (1957). *New Engl. J. Med. 256*, 1232

Meyers, F.H. (1972). In: *Review of medical pharmacology*, Lange Medical Publications, Los Altos, Calif.

Mezey, E. and Holt, P.R. (1971). *Exp. Molec. Pathol. 15*, 148

Miettinen, T.K. (1968). *Clin. Chim. Acta 20*, 43

Miller, E.C. (1978). *Cancer Res. 38*, 1479

Miller, E.C. Miller, J.A., Brown, R.R. and MacDonald, J.C. (1958). *Cancer Res. 18*, 469

Miller, J.A. and Miller, E.C. (1976). *Fed. Proc. 35*, 1316

Miranda, C.L., Mukhtar, H., Bend, J.R. and Chabra, R.S. (1979). *Biochem. Pharmacol. 28*, 2713

Miryish, S.S. (1971). *J. Natl Cancer Inst. 46*, 1183

Miryish, S.S., Cardesa, A., Wallcave, L. and Shubik, P. (1975). *J. Natl Cancer Inst. 55*, 633

Miryish, S.S., Wallcave, L., Eagen, M. and Shubik, P. (1972). *Science 177*, 65

Moir, A.T.B., Halliday, J. and Williams, I.R. (1971). *Lancet 2*, 798

Montgomery, R.D. (1965). *Am. J. Clin. Nutr. 17*, 103

Moore, R.B., Crane, C.A. and Frantz, I.D. (1968). *J. Clin. Invest. 47*, 1664

Moore, T. (1960). *Vit. Horm. 18*, 499

Moriarty, M.J., Mulgrew, S., Malone, J.R. and O'Connor, M.K. (1977). *Irish J. Med. Sci. 146*, 74

Munro, H.N. (1964). In: *Mammalian protein metabolism*, eds Munro, H.N. and Allison, J.B., vol. 1. Academic Press, New York, p. 381

Murphy, F.H. and Zelman, S. (1965). *J. Urol. 94*, 297

Muto, Y. and Moriwaki, H. (1984). *J. Natl Cancer Inst. 73*, 1389

Myerson, R.M. (1973). *Med. Clin. N. Am,. 57*, 925

Nala, G.T., Pope, S. and Harrison, D.C. (1970). *Am. Heart J. 79*, 499

Nalder, B.N., Mahoney, A.W., Ramakrishnan, R. and Hendricks, D.G. (1972). *J. Nutr. 102*, 535

Nandi, B.K., Majumder, A.K. and Halder, K. (1977). *Int. J. Vit. Nutr. Res. 47*, 200

Nandi, B.K., Majumder, A.K., Subramanian, N. and Chatterjee, I.B. (1973). *J. Nutr. 103*, 1688

Narisawa, T., Wong, C.Q., Maronpot, R.R. and Weisburger, J.H. (1976). *Cancer Res. 36*, 505

Natarajan, A.T., Tates, A.D. and Van Buul, P.O. (1976). *Mutat. Res. 37*, 83

Natoff, I.L. (1965). *Med. Bydraese (Netherlands) 11*, 101

Nayler, W.G. (1967). *Am. Heart J. 73*, 379

Nestel, P.J. and Hirsch, E.Z. (1965). *J. Lab. Clin. Med. 66*, 357

Nestel, P.J., Hirsch, E.Z. and Couzens, E.A. (1965). *J. Clin. Invest. 44*, 891

Nestel, P.J. and Whyte, H.M. (1968). *Metabolism 17*, 1122

Nettesheim, P., Snyder, C., Williams, M.L., Cone, M.V. and Kim, J.C. (1975). *Proc. Am. Assoc. Cancer Res. 16*, 54

Neuvonen, P.J. (1976). *Drugs 11*, 45

Neuvonen, P.J., Gothoni, G. and Hackman, R. (1970). *Br. Med. J. 4*, 532

Newberne, P.M. and Suphakan, V. (1977). *Cancer 40*, 2553

Newman, M.J.D. and Summon, D.W. (1957). *Blood 12*, 183

Newmark, H.L. and Mergens, W.J. (1981). In: *Inhibition of tumor production and development*, eds Zedeck, M.S. and Lipkins, M. Plenum Press, New York, p. 127

Newmark, H.L., Scheiner, M.S. and Marcus, M. (1976). *Am. J. Clin. Nutr. 29*, 645

Newmark, H.L., Scheiner, J. and Marcus, M. (1979) *J. Am. Med. Assoc. 242*, 2319

Nigro, N.D., Singh, D.V., Campbell, R.L. and Pak, M.S. (1975). *J. Natl Cancer Inst. 54*, 439

Nilsson, B.E. and Westlin, N.E. (1973). *Clin. Orthop. 90*, 229

Nomura, A.M.Y., Stemmermann, G.N., Heilbrun, L.K., Salkeld, R.M. and Vuillemier, J.P. (1985). *Cancer Res. 45*, 2369

Norred, W.P. and Wade, A.E. (1972). *Biochem. Pharmacol. 21*, 2887

Norris, J.W. and Pratt, R.F. (1971). *Neurol. 21*, 659

Nuessle, W.F. and Norman, F.C. (1965). *J. Am. Med. Assoc. 192*, 726

Olson, J.A. (1972). *Isr. J. Med. Sci. 8*, 1199

Olson, J.A. (1983). *Sem. Oncol. 10*, 290

O'Malley, K., Crooks, J., Duke, E. and Stevenson, I.H. (1971). *Br. Med. J. 2*, 607

Omura, H., Tomita, Y., Nakamura, Y. and Murakami, H. (1974). *J. Fac. Agric., Kyushu Univ. 18*, 181

Orfanos, C.E. and Schuppli, R. (1978). *Dermatologica 157* (Suppl. 1), 1

Osaki, S., McDermott, J.A. and Frieden, E. (1964). *J. Biol. Chem. 239*, 3570

Osborne, M.J., Freeman, M. and Huennekens, F.M. (1958). *Proc. Soc. Exp. Biol. Med. 97*, 429

O'Sullivan, J.B. and Mahan, C.M. (1964). *Diabetes 13*, 278

Oura, E., Konttinen, K. and Soumalainen, H. (1963). *Acta Physiol. Scand. 59*, 119

Ovesen, L. (1984). *Drugs 27*, 148

Paganini-Hill, A., Ross, R.K. and Gerkins, J.R. (1981). *Ann. Intern. Med. 95*, 28

Palva, J.P., Hemivarra, O. and Maltila, M. (1966). *Scand. J. Hematol. 3*, 149

Pamuken, A.M., Wattenberg, L.W., Price, J.M. and Bryan, G.T. (1971). *J. Natl Cancer Inst. 47*, 155

Pantuck, E.J., Hsiao, K.C., Conney, A.H., Garland, W., Kappas, A. and Anderson, K. (1976a). *Science 194*, 1055

Pantuck, E.J., Hsiao, K.C., Loub, W.D., Wattenberg, L.W., Kuntzman, R. and Conney, A.H. (1976b). *J. Pharmacol. Exp. Ther. 198*, 278

Parfitt, A.M., Gallagher, J.C. and Heany, R.P. (1982). *Am. J. Clin. Nutr. 36*, 1014

Parke, D.V. (1968). *The biochemistry of foreign compounds*. Pergamon Press, London, p. 172

Patel, J.M. and Pawan, S.S. (1973). *Ind. J. Biochem. Biophys. 10*, 73

Patel, J.M. and Pawan, S.S. (1974). *Biochem. Pharmacol. 23*, 1467

Paul, P.K. and Duttagupta, P.N. (1978). *Indian J. Exp. Biol. 16*, 18

Pawan, G.L.S. (1972). *Proc. Nutr. Soc. 31*, 83

Pawan, G.L.S. (1974). *Proc. Nutr. Soc. 33*, 239

Peers, F.G. and Linsell, C.A. (1973). *Br. J. Cancer 27*, 473

Pelletier, O. (1968). *Am. J. Clin. Nutr. 21*, 1259

Pelletier, O. (1975). *Ann. NY Acad. Sci. 258*, 156

Peloux, Y., Nofre, C., Cier, A. and Colbert, I. (1962). *Ann. Inst. Pasteur, Paris 102*, 6

Peraino, C., Fry, R.J.M. and Staffeldt, E. (1971). *Cancer Res. 32*, 1506

Perry, J. and Chanarin, I. (1972). *Gut 13*, 544

Peto, R., Doll, R., Buckley, J.D. and Sporn, M.B. (1981). *Nature (Lond.) 290*, 201

Pierides, A.M. (1981). *Drugs 21*, 241

Pierpaoli, P.G. (1972). *Drug. Intell. Clin. Pharm. 6*, 89

Pirola, R.C. and Lieber, C.S. (1972). *Pharmacology 7*, 185

Pochi, P.E. (1982). *Arch. Dermatol. 118*, 57

Pollack, E.S., Nomura, A.M., Heilbrun, L.K., Stemmermann, G.N. and Green, S.B. (1984). *New Engl. J. Med. 310*, 617

Pomare, E.W. and Heaton, K.W. (1973). *Br. Med. J. 4*, 262

Port, C.D., Sporn, M.B. and Kaufman, D.G. (1975). *Proc. Am. Assoc. Cancer Res. 16*, 21

Poser, E. and Smith, L.H. (1972). *New Engl. J. Med. 287*, 412

Poskitt, M.E. (1974). *Proc. Nutr Soc. 33*, 203

Prasad, A.S., Lei, K.Y., Moghissi, K.S., Stryker, J.C. and Oberleas, D. (1976). *Am. J. Obstet. Gynecol. 125*, 1063

Prasad, K.N. and Rama, B.N. (1984). In: *Vitamins, nutrition and cancer*, ed. Prasad, K.N., Karger, Basel, p. 76

Preece, J., Reynolds, E.H. and Johnson, A.L. (1971). *Epilepsia 12*, 335

Preston, R.S., Hayes, J.R., and Campbell, T.C. (1976). *Life Sci. 19*, 1191

Pryor, W.A. (1982). *Ann. NY Acad. Sci. 393*, 1

Race, T.F., Paes, I.C. and Faloon, W.W. (1970). *Am. J. Med. Sci. 259*, 32

Ralson, A.J., Hinley, J.B. and Snaith, R.P. (1970). *Lancet 1*, 867

Ravid, M. and Robson, M. (1976). *J. Am. Med. Assoc. 236* 1380

Recknagel, R.O. and Ghoshal, A.K. (1966). *Nature (Lond.) 210*, 1162

Reddy, B.S., Tanaka, T. and El-Bayoumy, K. (1985). *J. Natl Cancer Inst. 74*, 1325

Reddy, B.S., Weisburger, J.H. and Wynder, E.L. (1974). *Science 183*, 416

Reddy, B.S. and Wynder, E.L. (1975). *J. Nutr. 105*, 878

Reddy, B.S. and Wynder, E.L. (1973). *J. Natl Cancer Inst. 50*, 1437

Reidenberg, M.M. (1977). *Clin. Pharmacol. Ther. 22*, 729

151

Rettura, G., Duttagupta, C., Listowsky, P., Levenson, S.M. and Seifter, E. (1983). *Fed. Proc. 42*, 786

Rettura, G., Stratford, F., Levenson, S.M. and Seifter, E. (1982). *J. Natl Cancer Inst. 69*, 73

Reynolds,E.H. (1967). *Lancet 1*, 1086

Reynolds, E.H. (1973). *Lancet 1*, 1376

Reynolds, E.H. (1975). *Epilepsia 16*, 319

Reynolds, E.H. (1976). *Clinics Haematol. 5*, 661

Reynolds, E.H., Mattson, R.H. and Gallagher, B.B. (1971). *Neurol. 21*, 394

Reynolds, E.H. and Milner, G. (1966). *Quart. J. Med. 140*, 521

Rhodes, J., Stokes, P. and Abrams, P. (1984). *Cancer Immunol. Immunother. 16*, 189

Rice, S.L., Eitenmiller, R.R. and Koehler, P.E. (1976). *J. Milk Food Technol. 39*, 353

Richens, A. (1972). *Br. Med. J. 1*, 567

Richens, A. and Waters, A.H. (1971). *Br. J. Pharm. 41*, 414P

Richter, J.J. and Weiner, A. (1971). *J. Neurochem. 18*, 613

Ridges, A.P. (1973). In: *Biochemistry and mental illness,* eds Iversen, L.L. and Rose, S.P.R., Biochemical Society, London, p. 175

Rikans, L.E., Smith, C.R. and Zannoni, V.G. (1978), *J. Pharmacol. Exp. Ther. 204*, 702

Rippere, V. (1981). *Lancet 2*, 48

Rivers, J.M. (1975). *Am. J. Clin. Nutr. 28*, 550

Rodrigo, C., Antezana, C. and Baraona, E. (1971). *J. Nutr. 101*, 1307

Roe, D.A. (1974). *Life Sci. 15*, 1219

Roe, D.A. (1981). *J. Am. Diet. Assoc. 78*, 17

Roe, D.A. (1984). *Nutr. Rev. 42*, 141

Roepke, J.L.B. and Kirksey, A. (1978). *Fed. Proc. 37*, 449

Rogers, A.E. and Newberne, P.M. (1975). *Cancer Res. 35*, 3427

Rose, D.P. (1978). In: *Human vitamin B6 requirements,* eds National Research Council. National Academy of Sciences, Washington, DC, p. 193

Rose, D.P., Leklem, J.E., Fardal, L., Baron, R.B. and Shrago, E. (1977). *Am. J. Clin. Nutr. 30*, 691

Rose, D.P., Strong, R., Folkard, J. and Adams, P.W. (1973). *Am. J. Clin. Nutr. 26*, 48

Rosenberg, H.A. and Bates, R.T. (1976). *Clin. Pharmacol. Ther. 20*, 227

Rosenberg, I.H. and Goodwin, H.A. (1971). *Gastroenterology, 60*, 445

Rosenthal, G. (1971). *J. Am. Med. Assoc. 215*, 1671

Rowe, N.A. and Gorlin, R.J. (1959). *J. Dent. Res. 38*, 72

Rubin, E., Bacchin, P., Gang, H. and Lieber, C.S. (1970). *Lab Invest. 22*, 569

Rubin, E. and Lieber, C.S. (1971). *Science 172*, 1097

Russell, D.H. and Durie, G.M. (1978). *Polyamines as biochemical markers of normal and malignant growth.* Raven Press, New York

Rustia, M. (1975). *J. Natl Cancer Inst. 55*, 1389

Sackner, M.A. and Balian, L.J. (1960). *Am. J. Med. 28*, 135

Safer, D.J. and Allen, R.P. (1973). *Pediatrics 51*, 660

Saffiotti, U., Montesano, R., Sellakumar, A.R. and Borg, S. (1967). *Cancer Res. 20*, 857

Sahenk, Z. and Mendell, J.R. (1980). In: *Experimental and clinical neurotoxicology*, eds Spencer, P.S. and Schaumburg, H.H. Williams & Wilkins, Baltimore, p. 578

Salked, R.M., Knorr, K. and Korner, W.F. (1973). *Clin. Chim. Acta 49*, 195

Samborskaya, E.P. and Ferdman, T.D. (1966). *Bull. Exp. Biol. Med. 62*, 934

Samuel, P., Holtzman, C.M. and Goldstein, J. (1967). *Circulation 35*, 398

Sander, J., Schweinsberg, F. and Menz, H.P. (1968). *Hoppe Seylers Z. Physiol. Chem. 349*, 1691

Saroja, N., Millikarjuneswara, V.R. and Clemetson, C.A.B. (1971). *Contraception 3*, 269

Sato, P.H. and Zannoni, V.G. (1976). *J. Pharmacol. Exp. Ther. 198*, 295

Schaumburg, H., Kaplan, J., Windebank, A., Vick, N., Rasmus, S., Pleasure, D. and Brown, M.J. (1983). *New Engl. J. Med. 309*, 445

Schlegel, J.V. (1975). *Ann. NY Acad. Sci. 258*, 432

Schrauzer, G.N. (1984). In: *Vitamins, nutrition and cancer*, ed. Prasad, K.N., Karger, Basel, p. 240

Schrauzer, G.N. and Rhead, W.J. (1973). *Int. J. Vit. Nutr. Res. 43*, 201

Schwartz, J., Suda, D. and Light, G. (1986). *Biochem. Biophys. Res. Commun. 136*, 1130

Schwartz, J.F. (1961). *J. Am. Med. Assoc. 176*, 106

Sedaghat, A., Samuel, P., Crouse, J.R. and Ahrens, A.H. (1975). *J. Clin. Invest. 55*, 12

Seelig, M.S. (1969). *Ann. NY Acad. Sci. 147*, 537

Seifter, E., Rettura, G. and Levenson, S.M. (1984). *Fed. Proc. 43*, 662

Seifter, E., Rettura, G., Padawar, J. and Levenson, S.M. (1982). *J. Natl Cancer Inst. 68*, 835

Sellmeyer, E., Bhettay, E., Truswell, A.S., Meyers, O.L. and Hansen, J.D.L. (1972). *Arch. Dis. Child. 47*, 429

Seltzer, H.S. (1970). In: *Diabetes mellitus: theory and practice*, eds Ellenberg, M. and Rifkin, H. McGraw-Hill, New York, p. 346

Selvaraij, R.J. and Bhat, K.S. (1972). *Am. J. Clin. Nutr. 25*, 166

Selzer, A. and Cohn, K.E. (1970). *West. J. Med. 113*, 1

Sen, N.P. (1972). *Food Cosmet. Toxicol. 10*, 219

Sen, N.P., Smith, D.C., Ashwinghamer, L. and Howsam, B. (1970). *Can. Inst. Food Technol. J. 3*, 66

Shah, K.V., Barbhaiya, H.C. and Shrinivasan, V. (1968). *J. Ind. Med. Assoc. 51*, 127

Shamberger, R.J. (1970). *J. Natl Cancer Inst. 44*, 931

Shamberger, R.J., Rukovena, E., Longfield, A.K., Tytko, S.A., Deodhar, S. and Willis, C.E. (1973). *J. Natl Cancer Inst. 50*, 867

Shamberger, R.J. and Willis, C.E. (1971). *CRC Crit. Rev., Clin. Lab. Sci. 2*, 211

Shekelle, R.B., Lepper, M., Liu, S., Maliza, C., Raynor, W.J., Rossof, A.H., Paul, O., Shryock, A.M. and Stamler, J. (1981). *Lancet 2*, 1185

Shenefelt, R.E. (1972). *Am. J. Pathol. 66*, 589

Shively, C.A., Simons, R.J., Passananti, G.T., Dvorchik, B.H. and Vessell, E.S. (1981). *Clin. Pharmacol. Ther. 29*, 65

Shklar, G. (1982). *J. Natl Cancer Inst. 68*, 791

Shubik, P. (1975). *Cancer Res. 35*, 3475

Slater, T.F. (1966). *Nature (Lond.) 209*, 36

Smith, D.M., Rogers, A.E. and Newberne, P.M. (1975). *Cancer Res. 35*, 1485

Smith, E.D., Skalski, R.J., Johnson, G.C. and Ross, G.V. (1972). *J. Am. Med. Assoc. 221*, 1116

Smith, F.R. and Goodman, D.S. (1976). *New Engl. J. Med. 294*, 805

Smith, J.C., Brown, E.D., White, S.C. and Finkelstein, J.D. (1975). *Lancet 1*, 1251

Sodeman, W.A. (1965). *New Engl. J. Med. 273*, 93

Som, S., Chatterjee, M. and Banerjee, M.R. (1984). *Carcinogenesis 5*, 937

Sorensen, D.I., Devine, M. and River, J. (1974). *J. Nutr. 104*, 1041

Sorrell, M.F., Nauss, J.M., Donohue, T.M. and Tuma, D.J. (1983). *Gastroenterology 84*, 580

Spaans, F. (1970). *Epilepsia 11*, 403

Spellacy, W.M., Buhi, W.C. and Birk, S.A. (1972). *Contraception 6*, 265

Spivack, S.D. (1974). *Ann. Intern. Med. 81*, 795

Sporn, M.B., Dunlop, N.M., Newton, D.L. and Smith, J.M. (1976). *Fed. Proc. 35*, 1332

Sporn, M.B. and Roberts, A.B. (1984). *J. Natl Cancer Inst. 73*, 1381

Stamp, T.C.B. (1974). *Proc. R. Soc. Med. 67*, 64

Stein, H.B., Hasan, A. and Fox, I.H. (1976). *Ann. Intern. Med. 84*, 385

Stephen, J.M.L. (1968). *Br. J. Nutr. 22*, 153

Stoewsand, G.S., Broderick, B.D. and Bourke, J.B. (1970). In: *Pesticides symposia, Inter-American conference on toxicology and occupational medicine*, p. 139

Strisower, E.H. and Strisower, B. (1964). *J. Clin. Endocrinol. 24*, 139

Strobel, N.W., Lu, A.Y., Heidena, J. and Coon, M.J. (1970). *J. Biol. Chem. 245*, 4851

Strother, A., Throckmorton, J.K. and Herzer, C. (1971). *J. Pharmacol. Exp. Ther. 179*, 490

Suda, D., Schwartz, J. and Shklar, G. (1986). *Carcinogenesis 7*, 711

Sugerman, A.A. and Clark, C.G. (1974). *J. Am. Med. Assoc. 228*, 202

Surawicz, B. (1968). *Med. Clin. North Am. 52*, 1103

Sutton, J.L., Basu, T.K. and Dickerson, J.W.T. (1983). *Br. J. Nutr. 49*, 27

Swenson, D.H., Miller, E.C. and Miller, J.A. (1974). *Biochem. Biophys. Res. Commun. 60*, 1036

Swenson, D.H., Miller, J.A. and Miller, E.C. (1975). *Cancer Res. 35*, 3811

Szabo, A.J., Cole, H.S. and Grimaldi, R.D. (1970). *New Engl. J. Med. 19*, 646

Takiguchi, H., Furuyama, S. and Shimazono, N. (1966). *J. Vitaminol. 12*, 307

Talseth, T. (1977). *Clin. Pharmacokinet. 2*, 317

Tandler, B., Erlandson, R.A. and Wynder, E.L. (1968). *Am. J. Pathol. 52*, 69

Temple, N.J. and Basu, T.K. (1987). *J. Natl Cancer Inst. 78*, 1211

Tetzlaff, T.R., McCracken, G.H. Jr and Thomas, M.L. (1978). *J. Pediatr. 92*, 292

Thompson, H.J., Becci, P.J., Brown, C.C. and Moon, R.C. (1979). *Cancer Res. 39*, 3977

Thorp, V.J. (1980). *J. Am. Diet. Assoc. 76*, 581

Tomatis, I., Turusov, V. and Charles, R.T. (1972). *Int. J. Cancer 10*, 489

Tomkin, G.H., Hadden, D.R., Weaver, J.A. and Montgomery, D.A.D. (1971). *Br. Med. J. 2*, 685

Toth, B., Nagel, D., and Ross, A. (1982). *Br. J. Cancer 46*, 417

Tremolieres, J., Lowy, R. and Griffation, G. (1972). *Proc. Nutr. Soc. 31*, 107

Truswell, A.S. (1974). *Nutr. Soc. Proc. 33*, 215

Turley, S.D., West, C.E. and Horton, B.J. (1976). *Atherosclerosis 24*, 1

Tuyns, A.J. (1979). *Cancer Res. 39*, 2840

Ulland, B.M., Weisberger, J.H., Yamamoto, R.S. and Weisburger, E.K. (1973). *Food Cosmet. Toxicol. 11*, 199

Vakil, D.V., Ayiomannitis, A., Nizami, N. and Nizami, R.M. (1985). *Nutr. Res. 5*, 911

Van der Watt, J.J. (1974). In: *Mycotoxins,* ed. Purchase, I.F.H. Elsevier, Amsterdam, p. 369

Veitch, R.L., Lumeng, L. and Li, T.K. (1975). *J. Clin. Invest. 55*, 1026

Vernie, L.N. (1984). *Biochim. Biophys. Acta 738*, 203

Vernie, L.N., Bont, W.S., Ginjaar, H.B. and Emmelot, P. (1975). *Biochim. Biophys. Acta 414*, 283

Victor, M. and Adams, R.D. (1961). *Am. J. Clin. Nutr. 9*, 379

Visintine, R.F., Michaels, G.D., Fukayama, G., Conklin, J. and Kinsell, L.W. (1961). *Lancet 2*, 341

Wald, N.J., Boreham, J., Hayward, J.L. and Bulbrook, R.D. (1974). *Br. J. Cancer 49*, 321

Walker, A.I.T., Thorpe, E. and Stevenson, D.E. (1973). *Food Cosmet. Toxicol. 11*, 415

Walker, B.E., Kelleher, J., Dixon, M.F. and Losowsky, M.S. (1974). *Clin. Sci. Mol. Med. 47*, 449

Waller, J. (1978). *Accident Anal. Prev. 10*, 21

Wallerstein, R.O., Condit, P.K., Kasper, C.K., Brown, J.W. and Morrison, F.R. (1969). *J. Am. Med. Assoc. 208*, 2045

Walton, J.R. and Packer, L. (1980). In: *Vitamin E: a comprehensive treatise*, ed. Machlin, L.J. Marcel Dekker, New York, p. 495

Wasserman, R.H. (1970). In: *The fat soluble vitamins*, eds DeLuca, H.F. and Suttie, J.W. University of Wisconsin Press, Madison, Wis., p. 21

Waterlow, J.C. (1968). In: *Calorie deficiencies and protein deficiencies*, eds McCance, R.A. and Widdowson, E.M. Churchill, London, p. 61

Wattenberg, L.W. (1971). *Cancer 28*, 99

Wattenberg, L.W., Leong, J.L. and Strand, P.J. (1962). *Cancer Res. 22*, 1120

Wattenberg, L.W., Loub, W.D., Lam, L.K. and Speier, J.L. (1976). *Fed. Proc. 35*, 1327

Weatherholtz, W.M., Campbell, T.C. and Webb, R.E. (1969). *J. Nutr. 98*, 90

Weatherholtz, W.M. and Webb, R.E. (1971). *J. Nutr. 101*, 9

Webster, J.M. (1954). *Lancet 2*, 1017

Welling, P.G. (1977). *J. Pharmacokinet. Biopharm. 5*, 291

Welling, P.G., Huang, H., Koch, P.A., Craig, W.A. and Madsen, P.O. (1977). *J. Pharm. Sci. 66*, 549

Wells, D.G. (1968). *Lancet 1*, 146

Wentzler, R. (1979). In: *The vitamins*, Doubleday & Co., New York

Wharton, B.A. and McChesney, E.W. (1970). *J. Trop. Pediatr. 16*, 130

Wiener, S.G., Shoemaker, W.J., Doda, L.Y. and Bloom, F.E. (1981). *J. Pharmacol. Exp. Ther. 216*, 572

Wijnja, L. (1966). *Lancet 2*, 768

Wilkinson, P., Santamaria, J.N. and Rankin, J.G. (1969). *Aust. Ann. Med. 18*, 222

Williams, I.R. and Girdwood, R.H. (1970). *Scot. Med., J. 15*, 285

Willett, W.C., Polk, B.F., Underwood, B.A., Stampfer, M.J., Pressel, S., Rosner, B., Taylor, J.O., Schneider, K. and Hames, C.G. (1984). *New Engl. J. Med. 310*, 430

Wilson, C.W.M. and Loh, H.S. (1973). *Lancet 2*, 859

Windsor, A.C.M., Hobbs, C.B., Treby, D.A. and Astley-Cowper, R. (1972). *Br. Med. J. 1*, 214

Winston, F. (1973). *Am. J. Psychiat. 13*, 1217

Witting, L.A. (1974). *Am. J. Clin. Nutr. 27*, 952

Wogan, G.N. (1973). In: *Methods in cancer research*, vol. 7, ed. Busch, H. Academic Press, New York and London, p. 309

Wolf, G., Kiorpes, T.C. and Masushige, S. (1979). *Fed. Proc. 38*, 2540

Woodcock, B.G. and Wood, G.C. (1971). *Biochem. Pharmacol. 20*, 2703

Wu, A.H., Henderson, B.L., Pike, M.C. and Yu, M.C. (1985). *J. Natl Cancer Inst. 74*, 747

Wynder, E.L. (1975). *Cancer Res. 35*, 3388

Wynder, E.L. and Gori, G.P. (1977). *J. Natl Cancer Inst. 58*, 825

Wynn, V. and Doar, J.W.H. (1966). *Lancet 2*, 715

Wynn, V. and Doar, J.W.H. (1969). *Lancet 2*, 761

Yamafuji, K., Nakamura, Y., Omura, H., Soeda, T. and Gyotoku, K. (1971). *Z. Krebsforsch. 76*, 1

Yang, C.S. (1974). *Arch. Biochem. Biophys. 100*, 623

Yano, K., Rhoads, G. and Kaga, A. (1977). *New Engl. J. Med. 297*, 405

Yew, M.S. (1973). *Proc. Natl Acad. Sci. USA 70*, 969

Yosselson, S. (1976). *Drug. Intellig. Clin. Pharm. 10*, 8

Young, D.S., Pestaner, L.C. and Gibberman, V. (1975). *Clin. Chem. 21*, 1D

Yudkin, J. (1964). *Proc. Nutr. Soc. 23*, 149

Yunis, A.A., Arimura, G.K., Lutcher, C.L., Blasquez, J. and Halloran, M. (1967). *Blood 30*, 587

Zannoni, V.G. (1977). *Acta Vitam. Enzymol. (Milano) 32*, 17

Zannoni, V.G., Flynn, E.J. and Lynch, M. (1972). *Biochem. Pharmacol. 21*, 1377

Zannoni, V.G. and Lynch, M.M. (1973). *Drug Metabol. Rev. 2*, 57

Ziegler, R.G., Mason, J.J., Stemhagen, A., Hoover, R., Schoenberg, J.B., Gridley, G., Virgo, P.W., Altman, R. and Fraumeni, J.F. (1984). *J. Natl Cancer Inst. 73*, 1429

Zinn, M.B. (1970). *Tex. Med. 66*, 64

Index